Pitman Research Notes in Mathematics Series

Main Editors
H. Brezis, Université de Paris
R.G. Douglas, Texas A&M University
A. Jeffrey, University of Newcastle upon Tyne *(Founding Editor)*

Editorial Board
H. Amann, University of Zürich
R. Aris, University of Minnesota
G.I. Barenblatt, University of Cambridge
A. Bensoussan, INRIA, France
P. Bullen, University of British Columbia
S. Donaldson, University of Oxford
R.J. Elliott, University of Alberta
R.P. Gilbert, University of Delaware
D. Jerison, Massachusetts Institute of Technology
K. Kirchgässner, Universität Stuttgart
B. Lawson, State University of New York at Stony Brook
B. Moodie, University of Alberta
S. Mori, Kyoto University
L.E. Payne, Cornell University
G.F. Roach, University of Strathclyde
I. Stakgold, University of Delaware
W.A. Strauss, Brown University
S.J. Taylor, University of Virginia

Submission of proposals for consideration

Suggestions for publication, in the form of outlines and representative samples, are invited by the Editorial Board for assessment. Intending authors should approach one of the main editors or another member of the Editorial Board, citing the relevant AMS subject classifications. Alternatively, outlines may be sent directly to the publisher's offices. Refereeing is by members of the board and other mathematical authorities in the topic concerned, throughout the world.

Preparation of accepted manuscripts

On acceptance of a proposal, the publisher will supply full instructions for the preparation of manuscripts in a form suitable for direct photo-lithographic reproduction. Specially printed grid sheets can be provided and a contribution is offered by the publisher towards the cost of typing. Word processor output, subject to the publisher's approval, is also acceptable.

Illustrations should be prepared by the authors, ready for direct reproduction without further improvement. The use of hand-drawn symbols should be avoided wherever possible, in order to maintain maximum clarity of the text.

The publisher will be pleased to give any guidance necessary during the preparation of a typescript, and will be happy to answer any queries.

Important note

In order to avoid later retyping, intending authors are strongly urged not to begin final preparation of a typescript before receiving the publisher's guidelines. In this way it is hoped to preserve the uniform appearance of the series.

Addison Wesley Longman Ltd
Edinburgh Gate
Harlow, Essex, CM20 2JE
UK
(Telephone (0) 1279 623623)

UB MAGDEBURG   MA9
003 408 809

Titles in this series. A full list is available from the publisher on request.

100 Optimal control of variational inequalities
    V Barbu
101 Partial differential equations and dynamical systems
    W E Fitzgibbon III
102 Approximation of Hilbert space operators Volume II
    C Apostol, L A Fialkow, D A Herrero and D Voiculescu
103 Nondiscrete induction and iterative processes
    V Ptak and F-A Potra
104 Analytic functions – growth aspects
    O P Juneja and G P Kapoor
105 Theory of Tikhonov regularization for Fredholm equations of the first kind
    C W Groetsch
106 Nonlinear partial differential equations and free boundaries. Volume I
    J I Díaz
107 Tight and taut immersions of manifolds
    T E Cecil and P J Ryan
108 A layering method for viscous, incompressible L_p flows occupying R^n
    A Douglis and E B Fabes
109 Nonlinear partial differential equations and their applications: Collège de France Seminar. Volume VI
    H Brezis and J L Lions
110 Finite generalized quadrangles
    S E Payne and J A Thas
111 Advances in nonlinear waves. Volume II
    L Debnath
112 Topics in several complex variables
    E Ramírez de Arellano and D Sundararaman
113 Differential equations, flow invariance and applications
    N H Pavel
114 Geometrical combinatorics
    F C Holroyd and R J Wilson
115 Generators of strongly continuous semigroups
    J A van Casteren
116 Growth of algebras and Gelfand–Kirillov dimension
    G R Krause and T H Lenagan
117 Theory of bases and cones
    P K Kamthan and M Gupta
118 Linear groups and permutations
    A R Camina and E A Whelan
119 General Wiener–Hopf factorization methods
    F-O Speck
120 Free boundary problems: applications and theory. Volume III
    A Bossavit, A Damlamian and M Fremond
121 Free boundary problems: applications and theory. Volume IV
    A Bossavit, A Damlamian and M Fremond
122 Nonlinear partial differential equations and their applications: Collège de France Seminar. Volume VII
    H Brezis and J L Lions
123 Geometric methods in operator algebras
    H Araki and E G Effros
124 Infinite dimensional analysis–stochastic processes
    S Albeverio
125 Ennio de Giorgi Colloquium
    P Krée
126 Almost-periodic functions in abstract spaces
    S Zaidman
127 Nonlinear variational problems
    A Marino, L Modica, S Spagnolo and M Degliovanni
128 Second-order systems of partial differential equations in the plane
    L K Hua, W Lin and C-Q Wu
129 Asymptotics of high-order ordinary differential equations
    R B Paris and A D Wood
130 Stochastic differential equations
    R Wu
131 Differential geometry
    L A Cordero
132 Nonlinear differential equations
    J K Hale and P Martinez-Amores
133 Approximation theory and applications
    S P Singh
134 Near-rings and their links with groups
    J D P Meldrum
135 Estimating eigenvalues with *a posteriori/a priori* inequalities
    J R Kuttler and V G Sigillito
136 Regular semigroups as extensions
    F J Pastijn and M Petrich
137 Representations of rank one Lie groups
    D H Collingwood
138 Fractional calculus
    G F Roach and A C McBride
139 Hamilton's principle in continuum mechanics
    A Bedford
140 Numerical analysis
    D F Griffiths and G A Watson
141 Semigroups, theory and applications. Volume I
    H Brezis, M G Crandall and F Kappel
142 Distribution theorems of L-functions
    D Joyner
143 Recent developments in structured continua
    D De Kee and P Kaloni
144 Functional analysis and two-point differential operators
    J Locker
145 Numerical methods for partial differential equations
    S I Hariharan and T H Moulden
146 Completely bounded maps and dilations
    V I Paulsen
147 Harmonic analysis on the Heisenberg nilpotent Lie group
    W Schempp
148 Contributions to modern calculus of variations
    L Cesari
149 Nonlinear parabolic equations: qualitative properties of solutions
    L Boccardo and A Tesei
150 From local times to global geometry, control and physics
    K D Elworthy

151 A stochastic maximum principle for optimal control of diffusions
   U G Haussmann
152 Semigroups, theory and applications. Volume II
   H Brezis, M G Crandall and F Kappel
153 A general theory of integration in function spaces
   P Muldowney
154 Oakland Conference on partial differential equations and applied mathematics
   L R Bragg and J W Dettman
155 Contributions to nonlinear partial differential equations. Volume II
   J I Díaz and P L Lions
156 Semigroups of linear operators: an introduction
   A C McBride
157 Ordinary and partial differential equations
   B D Sleeman and R J Jarvis
158 Hyperbolic equations
   F Colombini and M K V Murthy
159 Linear topologies on a ring: an overview
   J S Golan
160 Dynamical systems and bifurcation theory
   M I Camacho, M J Pacifico and F Takens
161 Branched coverings and algebraic functions
   M Namba
162 Perturbation bounds for matrix eigenvalues
   R Bhatia
163 Defect minimization in operator equations: theory and applications
   R Reemtsen
164 Multidimensional Brownian excursions and potential theory
   K Burdzy
165 Viscosity solutions and optimal control
   R J Elliott
166 Nonlinear partial differential equations and their applications: Collège de France Seminar. Volume VIII
   H Brezis and J L Lions
167 Theory and applications of inverse problems
   H Haario
168 Energy stability and convection
   G P Galdi and B Straughan
169 Additive groups of rings. Volume II
   S Feigelstock
170 Numerical analysis 1987
   D F Griffiths and G A Watson
171 Surveys of some recent results in operator theory. Volume I
   J B Conway and B B Morrel
172 Amenable Banach algebras
   J-P Pier
173 Pseudo-orbits of contact forms
   A Bahri
174 Poisson algebras and Poisson manifolds
   K H Bhaskara and K Viswanath
175 Maximum principles and eigenvalue problems in partial differential equations
   P W Schaefer
176 Mathematical analysis of nonlinear, dynamic processes
   K U Grusa
177 Cordes' two-parameter spectral representation theory
   D F McGhee and R H Picard
178 Equivariant K-theory for proper actions
   N C Phillips
179 Elliptic operators, topology and asymptotic methods
   J Roe
180 Nonlinear evolution equations
   J K Engelbrecht, V E Fridman and E N Pelinovski
181 Nonlinear partial differential equations and their applications: Collège de France Seminar. Volume IX
   H Brezis and J L Lions
182 Critical points at infinity in some variational problems
   A Bahri
183 Recent developments in hyperbolic equations
   L Cattabriga, F Colombini, M K V Murthy and S Spagnolo
184 Optimization and identification of systems governed by evolution equations on Banach space
   N U Ahmed
185 Free boundary problems: theory and applications. Volume I
   K H Hoffmann and J Sprekels
186 Free boundary problems: theory and applications. Volume II
   K H Hoffmann and J Sprekels
187 An introduction to intersection homology theory
   F Kirwan
188 Derivatives, nuclei and dimensions on the frame of torsion theories
   J S Golan and H Simmons
189 Theory of reproducing kernels and its applications
   S Saitoh
190 Volterra integrodifferential equations in Banach spaces and applications
   G Da Prato and M Iannelli
191 Nest algebras
   K R Davidson
192 Surveys of some recent results in operator theory. Volume II
   J B Conway and B B Morrel
193 Nonlinear variational problems. Volume II
   A Marino and M K V Murthy
194 Stochastic processes with multidimensional parameter
   M E Dozzi
195 Prestressed bodies
   D Iesan
196 Hilbert space approach to some classical transforms
   R H Picard
197 Stochastic calculus in application
   J R Norris
198 Radical theory
   B J Gardner
199 The C^*-algebras of a class of solvable Lie groups
   X Wang
200 Stochastic analysis, path integration and dynamics
   K D Elworthy and J C Zambrini

201 Riemannian geometry and holonomy groups
   S Salamon
202 Strong asymptotics for extremal errors and polynomials associated with Erdös type weights
   D S Lubinsky
203 Optimal control of diffusion processes
   V S Borkar
204 Rings, modules and radicals
   B J Gardner
205 Two-parameter eigenvalue problems in ordinary differential equations
   M Faierman
206 Distributions and analytic functions
   R D Carmichael and D Mitrovic
207 Semicontinuity, relaxation and integral representation in the calculus of variations
   G Buttazzo
208 Recent advances in nonlinear elliptic and parabolic problems
   P Bénilan, M Chipot, L Evans and M Pierre
209 Model completions, ring representations and the topology of the Pierce sheaf
   A Carson
210 Retarded dynamical systems
   G Stepan
211 Function spaces, differential operators and nonlinear analysis
   L Paivarinta
212 Analytic function theory of one complex variable
   C C Yang, Y Komatu and K Niino
213 Elements of stability of visco-elastic fluids
   J Dunwoody
214 Jordan decomposition of generalized vector measures
   K D Schmidt
215 A mathematical analysis of bending of plates with transverse shear deformation
   C Constanda
216 Ordinary and partial differential equations. Volume II
   B D Sleeman and R J Jarvis
217 Hilbert modules over function algebras
   R G Douglas and V I Paulsen
218 Graph colourings
   R Wilson and R Nelson
219 Hardy-type inequalities
   A Kufner and B Opic
220 Nonlinear partial differential equations and their applications: Collège de France Seminar. Volume X
   H Brezis and J L Lions
221 Workshop on dynamical systems
   E Shiels and Z Coelho
222 Geometry and analysis in nonlinear dynamics
   H W Broer and F Takens
223 Fluid dynamical aspects of combustion theory
   M Onofri and A Tesei
224 Approximation of Hilbert space operators. Volume I. 2nd edition
   D Herrero
225 Operator theory: proceedings of the 1988 GPOTS–Wabash conference
   J B Conway and B B Morrel
226 Local cohomology and localization
   J L Bueso Montero, B Torrecillas Jover and A Verschoren
227 Nonlinear waves and dissipative effects
   D Fusco and A Jeffrey
228 Numerical analysis 1989
   D F Griffiths and G A Watson
229 Recent developments in structured continua. Volume II
   D De Kee and P Kaloni
230 Boolean methods in interpolation and approximation
   F J Delvos and W Schempp
231 Further advances in twistor theory. Volume I
   L J Mason and L P Hughston
232 Further advances in twistor theory. Volume II
   L J Mason, L P Hughston and P Z Kobak
233 Geometry in the neighborhood of invariant manifolds of maps and flows and linearization
   U Kirchgraber and K Palmer
234 Quantales and their applications
   K I Rosenthal
235 Integral equations and inverse problems
   V Petkov and R Lazarov
236 Pseudo-differential operators
   S R Simanca
237 A functional analytic approach to statistical experiments
   I M Bomze
238 Quantum mechanics, algebras and distributions
   D Dubin and M Hennings
239 Hamilton flows and evolution semigroups
   J Gzyl
240 Topics in controlled Markov chains
   V S Borkar
241 Invariant manifold theory for hydrodynamic transition
   S Sritharan
242 Lectures on the spectrum of $L^2(\Gamma\backslash G)$
   F L Williams
243 Progress in variational methods in Hamiltonian systems and elliptic equations
   M Girardi, M Matzeu and F Pacella
244 Optimization and nonlinear analysis
   A Ioffe, M Marcus and S Reich
245 Inverse problems and imaging
   G F Roach
246 Semigroup theory with applications to systems and control
   N U Ahmed
247 Periodic-parabolic boundary value problems and positivity
   P Hess
248 Distributions and pseudo-differential operators
   S Zaidman
249 Progress in partial differential equations: the Metz surveys
   M Chipot and J Saint Jean Paulin
250 Differential equations and control theory
   V Barbu

251 Stability of stochastic differential equations with respect to semimartingales
   X Mao
252 Fixed point theory and applications
   J Baillon and M Théra
253 Nonlinear hyperbolic equations and field theory
   M K V Murthy and S Spagnolo
254 Ordinary and partial differential equations. Volume III
   B D Sleeman and R J Jarvis
255 Harmonic maps into homogeneous spaces
   M Black
256 Boundary value and initial value problems in complex analysis: studies in complex analysis and its applications to PDEs 1
   R Kühnau and W Tutschke
257 Geometric function theory and applications of complex analysis in mechanics: studies in complex analysis and its applications to PDEs 2
   R Kühnau and W Tutschke
258 The development of statistics: recent contributions from China
   X R Chen, K T Fang and C C Yang
259 Multiplication of distributions and applications to partial differential equations
   M Oberguggenberger
260 Numerical analysis 1991
   D F Griffiths and G A Watson
261 Schur's algorithm and several applications
   M Bakonyi and T Constantinescu
262 Partial differential equations with complex analysis
   H Begehr and A Jeffrey
263 Partial differential equations with real analysis
   H Begehr and A Jeffrey
264 Solvability and bifurcations of nonlinear equations
   P Drábek
265 Orientational averaging in mechanics of solids
   A Lagzdins, V Tamuzs, G Teters and A Kregers
266 Progress in partial differential equations: elliptic and parabolic problems
   C Bandle, J Bemelmans, M Chipot, M Grüter and J Saint Jean Paulin
267 Progress in partial differential equations: calculus of variations, applications
   C Bandle, J Bemelmans, M Chipot, M Grüter and J Saint Jean Paulin
268 Stochastic partial differential equations and applications
   G Da Prato and L Tubaro
269 Partial differential equations and related subjects
   M Miranda
270 Operator algebras and topology
   W B Arveson, A S Mishchenko, M Putinar, M A Rieffel and S Stratila
271 Operator algebras and operator theory
   W B Arveson, A S Mishchenko, M Putinar, M A Rieffel and S Stratila
272 Ordinary and delay differential equations
   J Wiener and J K Hale
273 Partial differential equations
   J Wiener and J K Hale
274 Mathematical topics in fluid mechanics
   J F Rodrigues and A Sequeira
275 Green functions for second order parabolic integro-differential problems
   M G Garroni and J F Menaldi
276 Riemann waves and their applications
   M W Kalinowski
277 Banach C(K)-modules and operators preserving disjointness
   Y A Abramovich, E L Arenson and A K Kitover
278 Limit algebras: an introduction to subalgebras of C*-algebras
   S C Power
279 Abstract evolution equations, periodic problems and applications
   D Daners and P Koch Medina
280 Emerging applications in free boundary problems
   J Chadam and H Rasmussen
281 Free boundary problems involving solids
   J Chadam and H Rasmussen
282 Free boundary problems in fluid flow with applications
   J Chadam and H Rasmussen
283 Asymptotic problems in probability theory: stochastic models and diffusions on fractals
   K D Elworthy and N Ikeda
284 Asymptotic problems in probability theory: Wiener functionals and asymptotics
   K D Elworthy and N Ikeda
285 Dynamical systems
   R Bamon, R Labarca, J Lewowicz and J Palis
286 Models of hysteresis
   A Visintin
287 Moments in probability and approximation theory
   G A Anastassiou
288 Mathematical aspects of penetrative convection
   B Straughan
289 Ordinary and partial differential equations. Volume IV
   B D Sleeman and R J Jarvis
290 K-theory for real C^*-algebras
   H Schröder
291 Recent developments in theoretical fluid mechanics
   G P Galdi and J Necas
292 Propagation of a curved shock and nonlinear ray theory
   P Prasad
293 Non-classical elastic solids
   M Ciarletta and D Ieşan
294 Multigrid methods
   J Bramble
295 Entropy and partial differential equations
   W A Day
296 Progress in partial differential equations: the Metz surveys 2
   M Chipot
297 Nonstandard methods in the calculus of variations
   C Tuckey
298 Barrelledness, Baire-like- and (LF)-spaces
   M Kunzinger
299 Nonlinear partial differential equations and their applications. Collège de France Seminar. Volume XI
   H Brezis and J L Lions
300 Introduction to operator theory
   T Yoshino

301 Generalized fractional calculus and applications
   V Kiryakova
302 Nonlinear partial differential equations and their applications. Collège de France Seminar Volume XII
   H Brezis and J L Lions
303 Numerical analysis 1993
   D F Griffiths and G A Watson
304 Topics in abstract differential equations
   S Zaidman
305 Complex analysis and its applications
   C C Yang, G C Wen, K Y Li and Y M Chiang
306 Computational methods for fluid-structure interaction
   J M Crolet and R Ohayon
307 Random geometrically graph directed self-similar multifractals
   L Olsen
308 Progress in theoretical and computational fluid mechanics
   G P Galdi, J Málek and J Necas
309 Variational methods in Lorentzian geometry
   A Masiello
310 Stochastic analysis on infinite dimensional spaces
   H Kunita and H-H Kuo
311 Representations of Lie groups and quantum groups
   V Baldoni and M Picardello
312 Common zeros of polynomials in several variables and higher dimensional quadrature
   Y Xu
313 Extending modules
   N V Dung, D van Huynh, P F Smith and R Wisbauer
314 Progress in partial differential equations: the Metz surveys 3
   M Chipot, J Saint Jean Paulin and I Shafrir
315 Refined large deviation limit theorems
   V Vinogradov
316 Topological vector spaces, algebras and related areas
   A Lau and I Tweddle
317 Integral methods in science and engineering
   C Constanda
318 A method for computing unsteady flows in porous media
   R Raghavan and E Ozkan
319 Asymptotic theories for plates and shells
   R P Gilbert and K Hackl
320 Nonlinear variational problems and partial differential equations
   A Marino and M K V Murthy
321 Topics in abstract differential equations II
   S Zaidman
322 Diffraction by wedges
   B Budaev
323 Free boundary problems: theory and applications
   J I Diaz, M A Herrero, A Liñan and J L Vazquez
324 Recent developments in evolution equations
   A C McBride and G F Roach
325 Elliptic and parabolic problems: Pont-à-Mousson 1994
   C Bandle, J Bemelmans, M Chipot, J Saint Jean Paulin and I Shafrir
326 Calculus of variations, applications and computations: Pont-à-Mousson 1994
   C Bandle, J Bemelmans, M Chipot, J Saint Jean Paulin and I Shafrir
327 Conjugate gradient type methods for ill-posed problems
   M Hanke
328 A survey of preconditioned iterative methods
   A M Bruaset
329 A generalized Taylor's formula for functions of several variables and certain of its applications
   J-A Riestra
330 Semigroups of operators and spectral theory
   S Kantorovitz
331 Boundary-field equation methods for a class of nonlinear problems
   G N Gatica and G C Hsiao
332 Metrizable barrelled spaces
   J C Ferrando, M López Pellicer and L M Sánchez Ruiz
333 Real and complex singularities
   W L Marar
334 Hyperbolic sets, shadowing and persistence for noninvertible mappings in Banach spaces
   B Lani-Wayda
335 Nonlinear dynamics and pattern formation in the natural environment
   A Doelman and A van Harten
336 Developments in nonstandard mathematics
   N J Cutland, V Neves, F Oliveira and J Sousa-Pinto
337 Topological circle planes and topological quadrangles
   A E Schroth
338 Graph dynamics
   E Prisner
339 Localization and sheaves: a relative point of view
   P Jara, A Verschoren and C Vidal
340 Mathematical problems in semiconductor physics
   P Marcati, P A Markowich and R Natalini
341 Surveying a dynamical system: a study of the Gray–Scott reaction in a two-phase reactor
   K Alhumaizi and R Aris
342 Solution sets of differential equations in abstract spaces
   R Dragoni, J W Macki, P Nistri and P Zecca
343 Nonlinear partial differential equations
   A Benkirane and J-P Gossez

A Benkirane

Université Sidi Mohamed Ben Abdellah, Fés, Morocco

and

J-P Gossez

Université Libre de Bruxelles, Belgium

(Editors)

Nonlinear partial differential equations

(From a Conference in Fés)

 LONGMAN

Addison Wesley Longman Limited
Edinburgh Gate, Harlow
Essex CM20 2JE, England
and Associated Companies throughout the world.

*Published in the United States of America
by Addison Wesley Longman Inc.*

© Addison Wesley Longman Limited 1996

All rights reserved; no part of this publication may be reproduced, stored
in a retrieval system, or transmitted in any form or by any means,
electronic, mechanical, photocopying, recording, or otherwise, without
the prior written permission of the Publishers, or a licence permitting
restricted copying in the United Kingdom issued by the Copyright
Licensing Agency Ltd, 90 Tottenham Court Road, London, W1P 9HE

First published 1996

AMS Subject Classifications: (Main) 35J25
　　　　　　　　　　　　　　　(Subsidiary) 35J55, 35J70, 47H05

ISSN 0269-3674

ISBN 0 582 29213 1

British Library Cataloguing in Publication Data

　A catalogue record for this book is
　available from the British Library

Printed and bound in Great Britain
by Biddles Ltd, Guildford and King's Lynn

Contents

Preface

A Anane and N Tsouli
On the second eigenvalue of the p-Laplacian — 1

A Benkirane
A theorem of H Brezis and F E Browder type in Orlicz spaces and application — 10

K Benmlih
A result of maximum point localization for a semi-linear elliptic pde — 17

J Berkovits and V Mustonen
Existence results for semilinear equations with normal linear part and applications to telegraph equations — 32

L Boccardo
The role of truncates in nonlinear Dirichlet problems in L^1 — 42

A Boucherif and S M Bouguima
Periodic solutions of second ordinary differential equations with a discontinuous nonlinearity — 54

Z Chbani and H Riahi
The range of sums of monotone operators and applications to Hammerstein inclusions and nonlinear complementarity problems — 61

Ph Clement, D G de Figueiredo and E Mitidieri
A priori estimates for positive solutions of semilinear elliptic systems via Hardy–Sobolev inequalities — 73

B Dacorogna
On the minimisation of non quasiconvex integrals of the calculus of variations — 92

B Dehman
Local solvability for complex quasi-linear equations — 105

P Drábek
Strongly nonlinear degenerated and singular elliptic problems　　　　　　112

M Girardi and M Matzeu
Some results about periodic solutions of second order Hamiltonian systems where
the potential has indefinite sign　　　　　　　　　　　　　　　　　　　　　147

J P Gossez and M Moussaoui
A note on nonresonance between consecutive eigenvalues for a semilinear elliptic
problem　　　　　　　　　　　　　　　　　　　　　　　　　　　　　　　155

C A Stuart
Cylindrical TE and TM modes in a self-focusing dielectric　　　　　　　　167

J L Vazquez
Entropy solutions and the uniqueness problem for nonlinear second-order elliptic
equations　　　　　　　　　　　　　　　　　　　　　　　　　　　　　　179

Preface

The present collection of articles reflects some of the main subjects discussed at the "International Conference on Nonlinear Analysis" which was held in Fès, Morocco, on May 9-14, 1994. The goal of the Conference was to bring together a certain number of well-known specialists in nonlinear ordinary or partial differential equations so as to create some stimulating interaction with the Moroccan school of mathematical analysis. Several areas of the field were considered and are represented in this volume. More than half of the 15 papers included here deal with **nonlinear elliptic boundary value problems**, while the others are concerned with **monotone operator theory, calculus of variations, Hamiltonian systems**, ... Some of the papers are rather exhaustive surveys, while others contain deeply new results.

The Conference was organized by the "Faculté des Sciences Dhar Mahraz" from Fès. Financial support came from the "Université Sidi Mohamed Ben Abdellah", the "Société Mathématique du Maroc", the Sefrou Province, the ICTP and the International Mathematical Union. Many colleagues in Fès worked hard in the organization of the Conference, in particular Prof. A. Touzani. The final preparation of this volume largely benefited from the expertise of Mrs P. Leroy from the Department of Mathematics of the "Université Libre de Bruxelles".

It is a pleasure for us to thank all the people and institutions who contributed to the success of the Conference.

A. Benkirane,   J.-P. Gossez,

Fès.   Bruxelles.

A ANANE AND N TSOULI
On the second eigenvalue of the p-Laplacian

1 Introduction:

We consider the eigenvalue problem:

$$\begin{cases} -\Delta_p u = \lambda m(x)|u|^{p-2}u & in \ \Omega \\ u = 0 & on \ \partial\Omega. \end{cases} \quad (1)$$

where $\Delta_p = div(|\nabla u|^{p-2}\nabla u)$ is the p-Laplacian operator, Ω is a bounded domain of $\mathbb{R}^N$, $1 < p < \infty$ and $m \in L^\infty(\Omega)$ is a weight such that

$$mes(\Omega_m^+) \neq 0 \quad \text{with} \quad \Omega_m^+ = \{x \in \Omega : m(x) > 0\}. \quad (2)$$

Any nontrivial solution u, in the sense of distributions of (1), is said an eigenfunction associated to the eigenvalue λ.

Since 1984 many results have been obtained on the p-Laplacian spectrum, in particular on the first eigenvalue. In the case of N=1 and m=1, Otani [10] has explicited all the eigenvalues and has given a relatively complete description of the eigenfunctions when N> 1, Ω the ball B(0,R) and the weight m is positive and radial. De-Thélin [12] has shown the simplicity of λ_1 in the class of radial eigenfunctions.

Barles [4] and Sakaguchi [11] have established the simplicity of λ_1 when m=1 and $\partial\Omega$ is connected.

In the case of a regular domain and a weight m, it is established in [2] that the first eigenvalue of (1) is simple, isolated and is the only positive eigenvalue having an eigenfunction which does not change sign in Ω. It is also established in [3] that the problem (1) admits an unbounded sequence of positive eigenvalues (λ_n).

After this works, P. Lindquvist [8] has established the simplicity of λ_1 without any regularity assumption on Ω, he has also studied the continuity dependence of λ_1 on p [9].

Touzani [13] has considered an homogenous operator generalizing the p-Laplacian.

An other form of Δ_p defined by: $\Delta_p' u = \sum_{i=1}^N \frac{\partial}{\partial x_i}\left[\left|\frac{\partial u}{\partial x_i}\right|^{p-2}\frac{\partial u}{\partial x_i}\right]$ which is more degenerated than Δ_p has been considered recently by Benazzi [5]. W. Allegretto and Y.X. Huang [1] have established the simplicity of λ_1 when the domain Ω is the whole space $\mathbb{R}^N$. It is known in the case p=2 that if u_n is an eigenfunction associated to the eigenvalue λ_n, the number of components of the nodal set $\{x \in \Omega : u_n(x) \neq 0\}$ is not more than n (cf [6]). It is also known that an other important result in the case p=2 is the strict monotony dependence of λ_n with respect to the weight, i.e: If m, $m_0 \in L^\infty(\Omega)$ (two weights) such that $m_0 \leq \neq m$, with $mes(\Omega^+_{m_0}) \neq 0$ then

$$\lambda_n(\Omega, m_0) > \lambda_n(\Omega, m)$$

(cf [7]).

In this paper, we deal with the nodal property of the eigenfunctions of (1). We are particularly interested by the number of components of the nodal set defined by $\{x \in \Omega : u(x) \neq 0\}$. We prove analogous results to that of the well known case p=2 (cf [6]), this allows us to give a variational characterization of the second eigenvalue λ_2. We also establish the strict monotony dependence of λ_2 with respect to the weight.

2   Nodal eigenfunctions properties:

We are interested in weak solution of (1) i.e.

$$P(\Omega, m) \begin{cases} \text{find } u \in W_0^{1,p}(\Omega) \setminus \{0\} \quad \text{and } \lambda \in \mathbb{R} \quad \text{such that} \\ \int_\Omega |\nabla u|^{p-2} \nabla u . \nabla v \, dx = \lambda \int_\Omega m(x) |u|^{p-2} u . v \, dx, \quad v \in W_0^{1,p}(\Omega). \end{cases}$$

$\lambda \in \mathbb{R}$ is said an eigenvalue if there exists $u \in W_0^{1,p}(\Omega) \setminus \{0\}$ such that (u, λ) is solution of P(Ω,m), such a function u (nontrivial) is said an eigenfunction associated to λ.

We consider in $W_0^{1,p}(\Omega)$ the functionals:

$$A(v) = \frac{1}{p} \int_\Omega |\nabla v|^p \, dx,$$

$$B(v) = \frac{1}{p} \int_\Omega m |v|^p \, dx,$$

$$\Phi(v) = A(v)^2 - B(v).$$

It is clear that each critical point u (nontrivial) of Φ associated to a critical value c (i.e. $\Phi(u) = c$ and $\Phi'(u) = 0$) is an eigenfunction associated to the eigenvalue:

$$\lambda = \frac{1}{2\sqrt{-c}}.$$

Conversely, if $u \neq 0$ is an eigenfunction associated to a positive eigenvalue λ, $v = \left[(2\lambda A(u))^{-\frac{1}{p}} u \right]$ will be also an eigenfunction associated to $\lambda = \dfrac{1}{2A(v)}$ and v is a critical point of Φ associated to the critical value $c = -\dfrac{1}{4\lambda^2}$.

Let's consider the sequence $(c_n)_{n \in N}$ defined by:

$$c_n = \inf_{K \in A_n} \sup_{v \in K} \Phi(v), \tag{3}$$

where $A_n = \left\{ K \in W_0^{1,p}(\Omega) : K \text{ symmetrical compact and } \gamma(K) \geq n \right\}$ and $\gamma(K)$ denotes the genus of K. It is established in [3] that the sequence (c_n) is constituted by critical values of Φ tending to zero, then the sequence of eigenvalues (λ_n) defined by:

$$\lambda_n = \frac{1}{2\sqrt{-c_n}} \tag{4}$$

is positive nondecreasing tending to the infinity $(+\infty)$.

Let's denote by (u,λ) a solution of $P(\Omega, m)$, we define Z(u) by

$$Z(u) = \{ x \in \Omega : u(x) = 0 \}$$

and $N(u)$ is the number of components of $\Omega \setminus Z(u)$.

For each eigenfunction u associated to λ, we define $N(\lambda)$ by

$$N(\lambda) = \max \{ N(u) : (u, \lambda) \text{ solution of } P(\Omega, m) \},$$

in particular if $\lambda = \lambda_n$ for a certain integer n then we can write:

$$N(u) = N(\lambda).$$

Remark 1:
let (u, λ) be a solution of $P(\Omega, m)$, if ω is a component of $\Omega \setminus Z(u)$ then λ is the first eigenvalue of $P(\omega, m/_\omega)$, where $m/_\omega$ denotes the restriction of m to ω, the restriction of u to ω is then an associated eigenfunction and $\text{mes}(\Omega_m^+ \cap \omega) \neq 0$.

Remark 2:
Let ω be a component of $\Omega \setminus Z(u)$, we have:

$$mes(\omega) \geq (\lambda \|m\|_\infty c^p)^\tau = \beta > 0,$$

with c and τ are independent constants of ω, u and λ (cf[3]). Therefore

$$N(u) \leq \beta^{-1} mes(\Omega).$$

Proposition 1 *For each eigenvalue λ of $P(\Omega, m)$, we have :*

$$\lambda_{N(\lambda)} \leq \lambda.$$

Proof:

Let $r = N(\lambda)$, there exists an eigenfunction u associated to λ such that $N(u)=r$. Let $\omega_1, ..., \omega_r$ denote the r-components of $\Omega \setminus Z(u)$, for $i = 1, ..., r$ we set:

$$v_i = \begin{cases} \dfrac{u}{\left[\frac{1}{p}\int_{\omega_i} m\,|u|^p\,dx\right]^{\frac{1}{p}}} & in\ \omega_i \\ 0 & in\ \Omega \setminus \omega_i. \end{cases}$$

It is known that $v_i \in W_0^{1,p}(\Omega)$. Let F_r denotes the subspace of $W_0^{1,p}(\Omega)$ spanned by $\{v_1, ..., v_r\}$. For each $v \in F_r$, $v = \sum\limits_{i=1}^{r} \alpha_i v_i$, we have:

$$B(v) = \sum_{i=1}^{r} |\alpha_i|^p B(v_i) = \sum_{i=1}^{r} |\alpha_i|^p,$$

then the map $:v \longmapsto \left(B(v)^{\frac{1}{p}}\right)$ is a norme on F_r, therefore the compact K defined by:

$$K = \{v \in F_r : B(v) = -2c\}$$

where $c = -\dfrac{1}{4\lambda^2}$, can be identified to the sphere of $\mathbb{R}^r$, it has then genus r. By making $v = v_i$ in $P(\Omega, m)$, we obtain:

$$\int_{\omega_i} |\nabla v_i|^p\,dx = \lambda \int_{\omega_i} m(x)\,|v_i|^p\,dx \quad, for\ i = 1, ..., r.$$

So for $v \in K$, we have:

$$A(v) = \lambda \sum_{i=1}^{r} |\alpha_i|^p B(v_i) = \lambda \sum_{i=1}^{r} |\alpha_i|^p = \lambda B(v)$$

and
$$\Phi(v) = A(v)^2 - B(v) = \lambda^2 B(v)^2 - B(v) = c.$$

Since Φ is continuous and K is compact, then

$$c_r = \inf_{K \in A_r} \sup_{v \in K} \Phi(v) \leq \sup_{v \in K} \Phi(v) = c.$$

Therefore
$$\lambda_r \leq \lambda.$$

Corollar 1 *If $\lambda_n < \lambda_{n+1}$ then $N(n) \leq n$.*

Proof:

If $N(n) \geq n$ then $\lambda_{N(n)} \geq \lambda_{n+1}$, but by proposition 1:

$$\lambda_{N(n)} \leq \lambda_n,$$

so
$$\lambda_{n+1} = \lambda_n.$$

Remark 3:
From proposition 1, if q is the multiplicity order of λ_n (i.e. $\lambda_{n-1} < \lambda_n = \cdots = \lambda_{n+q} \leq \lambda_{n+q+1}$) then
$$N(n) \leq n + q,$$
we established in [14] that for each non null integer n, we have:
$$N(n) \leq 2n - 1.$$

3 On the second eigenvalue:

3.1 Variational Formulation of the second eigenvalue:

We shall give a variational formulation of the second eigenvalue of $P(\Omega, m)$. Precisely, we shall prove that the second eigenvalue of the p-Laplacian is the one obtained by (3):

Proposition 2 $\lambda_2 = \inf\{\lambda : \lambda \text{ positive eigenvalue of } P(\Omega, m) \text{ and } \lambda > \lambda_1\}$ *where λ_2 is given by (4).*

Proof:
Let $\mu = inf\{\lambda : \lambda \text{ positive eigenvalue of } P(\Omega,m) \text{ and } \lambda > \lambda_1\}$ since the set of eigenvalues of $P(\Omega,m)$ is closed ($cf[3]$) and λ_1 isolated, μ is a different eigenvalue of $P(\Omega,m)$ from λ_1. Therefore, if v is an eigenfunction associated to μ then the sign of v necessarily changes in Ω ($cf[3]$) i.e. $N(\mu) \geq 2$, which implies that:

$$\lambda_2 \leq \lambda_{N(\mu)}.$$

Proposition 1 provides, then
$$\lambda_{N(\mu)} \leq \mu.$$

So
$$\lambda_2 = \mu.$$

3.2 Monotone Dependence of λ_2 with respect to the weight:

It is easy to see that if m and m_0 are two weights in $L^\infty(\Omega)$ such that $m_0 \leq m$ and $\mathrm{mes}(\Omega_{m_0}^+) \neq 0$, then $\lambda_2(\Omega,m) \leq \lambda_2(\Omega,m_0)$. If $m_0 \lneq m$, it is well known in the case p=2 that the inequality is strict. In the general case, we have the following result:

Proposition 3 *Let m and m_0 be two weights in $L^\infty(\Omega)$ such that $m_0 \leq m$, $\mathrm{mes}(\Omega_{m_0}^+) \neq 0$ and $m_0 < m$ in $\Omega_{m_0}^+$ then*

$$\lambda_2(\Omega,m_0) > \lambda_2(\Omega,m).$$

Proof:
Let u_2 be an eigenfunction associated to $\lambda_2(\Omega,m_0)$, since its sign changes in Ω, we set:

$$v_1 = \begin{cases} \dfrac{u_2^+}{(\frac{1}{p}\int_{\Omega^+} m_0 \, |u_2^+|^p \, dx)^{\frac{1}{p}}} & \text{in } \Omega^+ \\ 0 & \text{in } \Omega \backslash \Omega^+ \end{cases}$$

$$v_2 = \begin{cases} \dfrac{u_2^-}{(\frac{1}{p}\int_{\Omega^-} m_0 \, |u_2^-|^p \, dx)^{\frac{1}{p}}} & \text{in } \Omega^- \\ 0 & \text{in } \Omega \backslash \Omega^- \end{cases}$$

where $\Omega^+ = \{x \in \Omega : u_2(x) > 0\}$ and $\Omega^- = \{x \in \Omega : u_2(x) < 0\}$.
It is clear that v_1 and v_2 are in $W_0^{1,p}(\Omega)$. Let $F_2 = span\{v_1, v_2\}$, the compact K defined by:

$$K = \{\, v \in F_2 : B_{m_0}(v) = -2c_2 \,\}$$

where $c_2 = \dfrac{-1}{4\lambda_2^2(\Omega, m_0)}$, has genus 2 (this was proved in proposition 1). By remark 1, we have :

$$mes(\Omega_{m_0}^+ \cap \Omega^+) \neq 0$$

and

$$mes(\Omega_{m_0}^- \cap \Omega^-) \neq 0.$$

Therefore

$$\int_{\Omega_{m_0}^+ \cap \Omega^+} m_0|v_i|^p \, dx < \int_{\Omega_{m_0}^+ \cap \Omega^+} m|v_i|^p \, dx \quad , \text{for i=1,2}$$

and

$$\int_{\Omega_{m_0}^- \cap \Omega^-} m_0|v_i|^p \, dx < \int_{\Omega_{m_0}^- \cap \Omega^-} m|v_i|^p \, dx \quad , \text{for i=1,2}.$$

Which implies

$$B_{m_0}(v_i) < B_m(v_i) \quad , \text{for i=1,2}.$$

Since for each v in K, we have:

$$\Phi_{m_0}(v) = A(v)^2 - B_{m_0}(v) = c_2,$$

so

$$\Phi_m(v) < \Phi_{m_0}(v) = c_2,$$

then

$$\sup_{v \in K} \Phi_m(v) < c_2,$$

therefore

$$c_2(\Omega, m) = \inf_{K \in A_n} \sup_{v \in K} \Phi_m(v) < c_2(\Omega, m_0).$$

Remark4:
Note that the above result of the strict monotony is partial. In the case $p = 2$, the proof uses essentially the fact that "inf-sup" in (3) is a "min-max".

Remark5:
It is established in [14] a similar result for higher eigenvalues.

Remark6:
An application of the results 2/ and 3/is used in [14] for studying the nonresonance for the Dirichlet problem:

$$\begin{cases} -\Delta_p u = f(x,u) + h(x) & \text{in } \Omega \\ u = 0 & \text{on } \partial\Omega, \end{cases}$$

when the nonlinearity f is between λ_1 and λ_2.

References

[1] W. Allegretto and Y.X. Huang, Eigenvalues of the indefinite weight p-Laplacian weigted $\mathbb{R}^n$ spaces, *Funkc.Ekerc*,(to appear).

[2] A. Anane, Simplicité et isolation de la première valeur propre du p-laplacien avec poids, *C.R.Ac.Paris*,305 (1987), 725-728.

[3] A. Anane, Etude des valeurs propres et de la résonance pour l'oprateur p-Laplacien, *Th.Doct.,U.L.B.*,1987.

[4] G. Barles, Remarks on uniqueness results of the first eigenvalue of the p-Laplacian,

[5] H. Benazzi, Etude du spectre d'un oprateur quasilinaire elliptique du second ordre, *Th.de troisime cycle., Universit.Mohammed.I.Oujda*, 1995.

[6] R. Courant and D. Hilbert, Methods of Mathematical Physics,VolumeII *Iterscience, New York*, 1962.

[7] D. De Figueiredo and J.P. Gossez, Strict monotonicity of eigenvalues and unique continuation, Comm.Par.Diff.Eq.,17(1992), 339-346.

[8] P. Lindqvist, On the equation $div(|\nabla u|^{p-2}\nabla u) + \lambda |u|^{p-2} u = 0$, Proc. of Amer.Math.Soc.,109 (1) (1990).

[9] P. Lindqvist, On a nonlinear eigenvalue problem: stability and concavity , *Research repport*, A279, June 1990, Helsinki Univ. Techn. Inst .Mathem .

[10] M. Ôtani, A remark on certain nonlinear elliptic equations , *Proc. Fac. Sci. Tokai Univ.*, 19 (1984), 23-28.

[11] S. Sakaguchi, Concavity properties of solutions to some degenerate quasilinear elliptic Dirichlet problems,*Annali della Scuala Normale Superiore de Pisa,Serie IV*, (Classe di Scienze) 14 (1987), 403-421.

[12] F. de Thélin, Sur l'espace propre associé la première valeur propre du pseudo-Laplacien, *C.R.Acad.Sc.Paris*, t.303, Série I N 8 (1986), 355-358.

[13] A. Touzani, Quelques résultats sur le A_p-Laplacien avec poids indéfini. *Th.Doct.*, U.L.B., 1990

[14] N. Tsouli, Etude de l'ensemble nodale des fonctions propres et de la non-résonance pour l'operateur p-Laplacien. *Th.Doct.* Universit. Mohammed Ier. Oujda.

Département de Mathématiques - Faculté des sciences - Oujda - Maroc

A BENKIRANE

A theorem of H Brezis and F E Browder type in Orlicz spaces and application

1 Introduction

Let Ω be an open subset of $\mathbf{R}^n$, $m \in \mathbf{N}$ and $1 < p, p' < +\infty$ such that $\frac{1}{p} + \frac{1}{p'} = 1$. Consider u in $W_0^{m,p}(\Omega), u \geq 0$ a.e. in Ω and T in $W^{-m,p'}(\Omega)$. Assume that $T = \mu + h$, where μ is a positive Radon measure and h an $L^1_{loc}(\Omega)$ function. L. Boccardo, D. Giachetti and F. Murat [11] proved that if the following condition is satisfied:

$$h(x)u(x) \geq -|\Phi(x)| \quad \text{a.e. } x \in \Omega \quad \text{for some } \Phi \text{ in } L^1(\Omega).$$

Then:

$$hu \in L^1(\Omega), u \in L^1(\Omega, d\mu) \text{ and } \quad \langle T, u \rangle = \int_\Omega u d\mu + \int_\Omega h u dx \qquad (1\text{-}1)$$

This result, which extends previous theorems of H. Brézis and F.E. Browder [12] where $\mu = 0$ or $h = 0$, is applied in [11] to resolve the following non linear variational inequality:

$$u \in K_\Psi, \; g(.,u) \in L^1(\Omega), \; ug(.,u) \in L^1(\Omega)$$

$$< Au, v - u > + \int_\Omega g(.,u)(v-u)dx \; \geq \; < f, v - u > \qquad (1\text{-}2)$$

$$\forall v \in K_\Psi \cap L^\infty(\Omega)$$

Where A is a pseudo-monotone operator acting on $W_0^{m,p}(\Omega)$, $f \in W^{-m,p'}(\Omega)$, $K_\Psi = \{v : v \in W_0^{m,p}(\Omega), v \geq \Psi \text{ a.e. in } \Omega\}$, $\Psi \in W_0^{m,p}(\Omega) \cap L^\infty(\Omega)$, and g satisfies the sign condition $sg(x,s) \geq 0$ but no growth restriction with respect to s.

It is our purpose in this paper, to study the problems (1-1) and (1-2) in the setting of the Orlicz-Sobolev spaces $W^m L_M(\Omega)$. Our results are obtained under the assumption that both the N-function M and its conjugate $\overline{M}$ satisfy the Δ_2 condition. These conditions intervene essentially in the use of some tools of non linear potential theory in Orlicz spaces, as Hedberg's approximation or properties of capacity. It is not clear whether the present approach can be further adapted to obtain the results for general N-functions (see [3], [4] and [10]).

2 Preliminaries

2.1. In this subsection we list briefly some definitions and well-known facts about Orlicz-spaces and Orlicz-Sobolev spaces. Standard references are [1], [17], [18]. Let $M : \mathbf{R}^+ \to \mathbf{R}^+$ be an N-function, i.e. M is continuous, convex, with $M(t) > 0$ for $t > 0$, $\frac{M(t)}{t} \to 0$ as $t \to 0$ and $\frac{M(t)}{t} \to \infty$ as $t \to \infty$.

Equivalently, M admits the representation: $M(t) = \int_0^t a(\tau)d\tau$
where $a : \mathbf{R}^+ \to \mathbf{R}^+$ is non-decreasing, right continuous, with $a(0) = 0$, $a(t) > 0$ for $t > 0$ and $a(t) \to \infty$ as $t \to \infty$.

The N-function $\overline{M}$ conjugate to M is defined by $\overline{M}(t) = \int_0^t \overline{a}(\tau)d\tau$, where $\overline{a} : \mathbf{R}^+ \to \mathbf{R}^+$ is given by $\overline{a}(t) = \sup\{s : a(s) \leq t\}$.

The N-function M is said to satisfy the Δ_2 condition if, for some k:
$$M(2t) \leq k\, M(t) \qquad \forall t \geq 0$$
We will extend these N-functions into even functions on all $\mathbf{R}$.

Let Ω be an open subset of $\mathbf{R}^n$. The Orlicz space $L_M(\Omega)$ is defined as the set of (equivalence classes of) real-valued measurable functions u on Ω such that:
$$\int_\Omega M\left(\frac{u(x)}{\lambda}\right)dx < +\infty \quad \text{for some } \lambda > 0$$

$L_M(\Omega)$ is a Banach space under the norm:
$$\|u\|_{M,\Omega} = \inf\left\{\lambda > 0 : \int_\Omega M\left(\frac{u(x)}{\lambda}\right)dx \leq 1\right\}$$

Let M and N be two N-functions, we say that A and B are equivalents (resp. near infinity) if there exist $k_1 > k_2 > 0$ such that:
$$M(k_1 t) \geq N(t) \geq M(k_2 t) \;\; \forall t \in \mathbf{R}^+ \;\; (\text{resp.}\forall t \geq t_0 > 0)$$

This implies the equality $L_M(\Omega) = L_N(\Omega)$ for all (resp. bounded) open subset of $\mathbf{R}^n$.

The closure in $L_M(\Omega)$ of the set of bounded measurable functions with compact support in $\overline{\Omega}$ is denoted by $E_M(\Omega)$. The equality $E_M(\Omega) = L_M(\Omega)$ holds if and only if M satisfies the Δ_2 condition, for all t or for t large according to whether Ω has infinite measure or not.

The dual of $E_M(\Omega)$ can be identified with $L_{\overline{M}}(\Omega)$ by means of the pairing $\int_\Omega u(x)v(x)dx$, as the dual norm on $L_{\overline{M}}(\Omega)$ is equivalent to $\|.\|_{\overline{M},\Omega}$. The space $L_M(\Omega)$ is reflexive if and only if M and $\overline{M}$ satisfy the Δ_2 condition, for all t or for t large, according to whether Ω has infinite measure or not.

We now turn to the Orlicz-Sobolev space. $W^m L_M(\Omega)$ (resp. $W^m E_M(\Omega)$) is the space of all functions u such that u and its distributional derivatives up to order m lie in $L_M(\Omega)$ (resp. $E_M(\Omega)$).

It is a Banach space under the norm:

$$\|u\|_{m,M,\Omega} = \sum_{|\alpha|\leq m} \|D^\alpha u\|_{M,\Omega}$$

Thus $W^m L_M(\Omega)$ and $W^m E_M(\Omega)$ can be identified with subspaces of the product of copies of $L_M(\Omega)$ (resp.$E_M(\Omega)$). Denoting this product by ΠL_M (resp.ΠE_M), we will use (in these preliminaries) the weak topologies $\sigma(\Pi L_M, \Pi E_{\overline{M}})$ and $\sigma(\Pi L_M, \Pi L_{\overline{M}})$.

The space $W_0^m E_M(\Omega)$ is defined as the (norm) closure of the Schwartz space $D(\Omega)$ in $W^m E_M(\Omega)$ and the space $W_0^m L_M(\Omega)$ as the $\sigma(\Pi L_M, \Pi E_{\overline{M}})$ closure of $D(\Omega)$ in $W^m L_M(\Omega)$. Let $W^{-m} L_{\overline{M}}(\Omega)$ (resp. $W^{-m} E_{\overline{M}}(\Omega)$) denote the space of distributions on Ω which can be written as sums of derivatives of order $\leq m$ of functions in $L_{\overline{M}}(\Omega)$ (resp. $E_{\overline{M}}(\Omega)$). It is a Banach space under the usual quotient norm. If the open set Ω has the segment property, then the space $D(\Omega)$ is dense in $W_0^m L_M(\Omega)$ for the topology $\sigma(\Pi L_M, \Pi L_{\overline{M}})$ (cf.[14],[15]). Consequently, the action of a distribution in $W^{-m} L_{\overline{M}}(\Omega)$ on an element of $W_0^m L_M(\Omega)$ is well defined.

2.2. We recall now the Orlicz-Sobolev imbedding theorem of Adams, Donaldson &Trudinger ([2] [13]). Let M be an N-function and N an N-function which is equivalent to M near infinity and such that:

$$\int_0^1 \frac{N^{-1}(t)}{t^{1+\frac{1}{n}}} dt < +\infty$$

If $\int_1^{+\infty} \frac{N^{-1}(t)}{t^{1+\frac{1}{n}}} dt = +\infty$, then we define a new N-function N_1 by the formula:

$$N_1^{-1}(s) = \int_0^s \frac{N^{-1}(t)}{t^{1+\frac{1}{n}}} dt$$

Let now M_1 be an N-function which is equivalent to N_1 near infinity and which is equal to M near zero. Repeating this process, one obtains a finite sequence of N-functions $M_0 = M$, $M_1, M_2 = (M_1)_1 ... M_q = (M_{q-1})_1$. Where $q = q(M,n)$ is such that : $\int_1^{+\infty} \frac{M_{q-1}^{-1}(t)}{t^{1+\frac{1}{n}}} dt = +\infty$ and $\int_1^{+\infty} \frac{M_q^{-1}(t)}{t^{1+\frac{1}{n}}} dt < +\infty$. Let Ω be an open subset of $\mathbf{R}^n$ with cone property, if $m \leq q(M,n)$ (resp. $m > q(M,n)$), then $W^m L_M(\Omega) \subset L_{M_m}(\Omega)$ (resp. $W^m L_M(\Omega) \subset C(\Omega) \cap L^\infty(\Omega)$) with continuous injection.

2.3. We close this section by a brief summary of some results about capacity in Orlicz spaces (for more details about potential theory in Orlicz spaces, see [3] [4] [5]). Let M be an N-function and m be a positive integer. If $X \subset \mathbf{R}^n$, we define the (m, M)-capacity of X by the formula:

$$C_{m,M}(X) = \inf \{\|u\|_M, u \in L_M(\mathbf{R}^n), g_m * u \geq 1 \text{ on } X\}$$

where g_m denote the usual Bessel function of order m.

It is known that $C_{m,M}$ is an outer capacity defined on all subsets of $\mathbf{R}^n$. If a statement is true for all x except those belonging to a set E with $C_{m,M}(E) = 0$, we say that it is true quasieverywhere (q.e.).

Lemma 1:
Let M be an N-function such that M and $\overline{M}$ satisfy the Δ_2 condition, and let $u \in W^m L_M(\mathbf{R}^n)$, then:

(i) u has a (unique) quasi-continuous representative. Moreover $u \geq 0$ a.e. if and only if $u \geq 0$ q.e.

(ii) If u_k is a sequence which converge to u in $W^m L_M(\mathbf{R}^n)$, then there exist a subsequence of u_k which converge q.e. to u.

Lemma 2:
Let M be an N-function such that M and $\overline{M}$ satisfy the Δ_2 condition, and let Ω be an open subset of $\mathbf{R}^n$. Let $m \in \mathbf{N}$ such that $m \leq q(M, n)$ and $T \in W^{-m} L_{\overline{M}}(\Omega) \cap \mathsf{M}(\Omega)$, where $\mathsf{M}(\Omega)$ denote the set of Radon measures on Ω. If $X \subset \Omega$ is such that $C_{m,M}(X) = 0$, then X is $|T|$-mesurable and $|T|(X) = 0$.

Let us point out that if $m > q(M, n)$, then $C_{m,M}(X) \neq 0$ for all $X \subset \mathbf{R}^n$.

3 A theorem of Brézis & Browder type

In this section we study the following question : let $u \in W_0^m L_M(\Omega)$ and $T \in W^{-m} L_{\overline{M}}(\Omega)$ such that $T = \mu + h$, where μ lies in $\mathsf{M}^+(\Omega)$ (the subset of positive Radon measures) and h in $L^1_{loc}(\Omega)$; find sufficient conditions on the data in order for u to belong to $L^1(\Omega; d\mu)$, for hu to belong to $L^1(\Omega)$ and finally to have:
$$<T, u> = \int_\Omega u d\mu + \int_\Omega hu dx.$$

Remark that even if all these expressions make sense, it is not obvious that the equality holds true. This question was solved in [10] when $\mu = 0$ and in [11] in the case of the classical Sobolev spaces. In [10] the N-function and its conjugate are supposed to satisfy the Δ_2 condition near infinity and the main tool in the proof is an approximation theorem of Hedberg type in $W^m L_M(\Omega)$ (see also [6]).

Our result is the following:

Theorem 1:
Let M be an N-function such that M and $\overline{M}$ satisfy the Δ_2 condition, and let Ω be an open subset of $\mathbf{R}^n$ and $m \in \mathbf{N}$. Consider $u \in W_0^m L_M(\Omega)$, $u \geq 0$ a.e. in Ω and $T \in W^{-m} L_{\overline{M}}(\Omega)$ such that $T = \mu + h$, where $\mu \in \mathsf{M}^+(\Omega)$ and $h \in L^1_{loc}(\Omega)$, assume that :

$$hu \geq -|\Phi| \text{ a.e. in } \Omega \text{ for some } \Phi \text{ in } L^1(\Omega) \qquad (3-1)$$

Then:

$hu \in L^1(\Omega)$, $u \in L^1(\Omega; d\mu)$ and $<T, u> = \int_\Omega u d\mu + \int_\Omega hu dx$ \hfill $(3-2)$

Remark 1:
Note that $\mu(F) = 0$ for all $F \subset \Omega$ such that $C_{m,M}(F) = 0$. Indeed, by lemma2: $|T|(F) = |\mu + h|(F) = 0$ but $0 \leq \mu(F) \leq |h|(F) + |\mu + h|(F) = 0$.

Proof of theorem1:
Some arguments are easily adapted from [10] and [11], and we will only sketch them. It is easy to see that we can assume without loss of generality that u has compact support in $\overline{\Omega}$. Indeed, take $\zeta \in \mathbf{D}(\mathbf{R}^n)$ with $0 \leq \zeta \leq 1$, $\zeta = 0$ outside the unit ball and $\zeta = 1$ in a neighbourhoud of zero. Put $\zeta_l(x) = \zeta(\frac{x}{l})$ and define $v_l = u\zeta_l$, then $v_l \to u$ as $l \to +\infty$ for the norm of $W_0^m L_M(\Omega)$ (also a.e.; q.e. and μ-a.e., for a subsequence, by lemmas 1 and 2 and by remark1).

If we can prove that:

$hv_l \in L^1(\Omega)$, $v_l \in L^1(\Omega; d\mu)$ and $<T, v_l> = \int_\Omega v_l d\mu + \int_\Omega hv_l dx$

Then the conclusion follows as in [11] by letting $l \to +\infty$ and by using $(3-1)$.

Consider first the case $m > q(M, n)$. Then $u \in L^\infty(\Omega)$, and the proof of Theorem 4 in [15] yields the existence of a sequence $\varphi_j \in \mathbf{D}(\mathbf{R}^n)$ such that $\varphi_j \to u$ for the norm of $W_0^m L_M(\Omega)$ as well a.e. and q.e. in Ω, $|\varphi_j(x)| \leq$ some constant a.e. in Ω and supp $\varphi_j \subset$ a fixed compact set $\subset \overline{\Omega}$. Using this sequence, one easily obtains $(3-2)$.

Consider now the case $m \leq q(M, n)$. We extend u by zero outside Ω and denote by $\tilde{u}$ the resulting function. Since $\tilde{u} \in W^m L_M(\mathbf{R}^n)$, there exist $f \in L_M(\mathbf{R}^n)$ such that (see[10]):

$\tilde{u} = I_m * f$, where I_m is the Riesz potential of order m

Put $w = I_m * |f|$ and $u_k(x) = H(\frac{w(x)}{k}) \tilde{u}(x)$ where $H \in C^\infty(\mathbf{R})$ such that $0 \leq H \leq 1$ and $H = 1$ on $[\frac{-1}{2}, \frac{1}{2}]$ and 0 outside $[-1, 1]$. We have $u_k \to \tilde{u}$ in $W^m L_M(\mathbf{R}^n)$ (see [10]).

Let $\varphi_j \in \mathbf{D}(\Omega)$ such that $\varphi_j \to u$ for the norm of $W_0^m L_M(\Omega)$ as well a.e. and q.e. in Ω. For each j we perform the same construction:

$\widetilde{\varphi_j} = I_m * f_j$ with $f_j \in L_M(\mathbf{R}^n)$; $w_j = I_m * |f_j|$ and $\varphi_{j,k}(x) = H(\frac{w_j(x)}{k}) \widetilde{\varphi_j}(x)$.

Since $\varphi_{j,k} \in W_0^m L_M(\Omega) \cap L^\infty(\Omega)$ has compact support in Ω, a simple mollification yields:

$<T, \varphi_{j,k}> = \int_\Omega \varphi_{j,k} d\mu + \int_\Omega h\varphi_{j,k} dx$ \hfill $(3-4)$

Now, by the well known properties of Riesz's potential (see theorem3.9 of [10]):

$\varphi_{j,k} \to u_k$ a.e. in Ω and $D^\alpha \varphi_{j,k} \to D^\alpha u_k$ a.e. in Ω.

and by the usuals estimates of $D^\alpha \varphi_{j,k}$ (see[10]) , an application of Vitali's theorem shows that $\varphi_{j,k} \to u_k$ in $W^m L_M(\mathbf{R}^n)$, a.e. and μ-a.e. in Ω.

Then, since $|\varphi_{j,k}| \leq k$, Lebesgue's theorem gives by letting $j \to \infty$ in $(3-4)$:

$$< T, u_k > = \int_\Omega u_k d\mu + \int_\Omega h u_k dx \qquad (3-5)$$

Finally, letting $k \to \infty$ in $(3-5)$, we deduce $(3-2)$ as in $[11]$.

4   Application to an unilateral problem

Let M be an N-function such that M and $\overline{M}$ satisfy the Δ_2 condition and Ω be an open subset of $\mathbf{R}^n$, $m \in \mathbf{N}$ .We consider a pseudo-monotone mapping S from $W_0^m L_M(\Omega)$ into $W^{-m} L_{\overline{M}}(\Omega)$ which satisfies the following conditions:

(i) S is continuous from each finite-dimensional subspace of $W_0^m L_M(\Omega)$ into $W^{-m} L_{\overline{M}}(\Omega)$ for the weak* topologie.

(ii) Su remains bounded in $W^{-m} L_{\overline{M}}(\Omega)$ whehever u remains bounded in $W_0^m L_M(\Omega)$ and $< u, Su >$ remains bounded from above.

(iii) For any given $f_0 \in W_0^m L_M(\Omega)$, $< u, Su - f_0 >> 0$ when u has sufficiently large norm.

Let $g : \Omega \times \mathbf{R} \to \mathbf{R}$ be a carathéodory function such that:

(a) $s.g(x,s) \geq 0 \; \forall s \in \mathbf{R}$ et a.e. in Ω.

(b) $h_t(x) = \sup_{|s| \leq t} |g(x,s)| \in L^1(\Omega) \quad \forall t \in \mathbf{R}^+$.

Finally consider some right hand side $f \in W^{-m} L_{\overline{M}}(\Omega)$ and the convex set $K_\Psi = \{v : v \in W_0^m L_M(\Omega), v \geq \Psi \text{ a.e. in } \Omega\}$ where the obstacle Ψ is assumed to belong to $W_0^m L_M(\Omega) \cap L^\infty(\Omega)$.

Theorem 2:

The variational inequality:

$$u \in K_\Psi \, , \, g(.,u) \in L^1(\Omega) \, , \, ug(.,u) \in L^1(\Omega)$$

$$< Su, v - u > + \int_\Omega g(.,u)(v-u)dx \geq < f, v - u > , \quad \forall v \in K_\Psi \cap L^\infty(\Omega)$$

has at least one solution.

Proof:

It is easily adapted from that given in [11] in the case of the classical Sobolev spaces, by using proposition1 of [16] and our theorem1, in place of the classical Lions's existence theorem and theorem 1-3 of [11].

Remark 2:

In reference [16], this last theorem is proved in the case $m = 1$ for general N-functions M (see also [8] and [9] for the case of strongly non linear equations).

References

[1] *R. Adams*, Sobolev spaces, Ac. Press , NewYork,1975.

[2] *R. Adams*, On the Orlicz-Sobolev imbedding theorem, J. Fun. Analysis, 24 (1977) ,241-257.

[3] *N. Aïssaoui*, une théorie du potentiel dans les espaces d'Orlicz , Doctorat d'état Université de Fès, 1994.

[4]*N.Aïssaoui et A. Benkirane,* Capacités dans les espaces d'Orlicz, Ann. Sci. Math. Québec 18 (1994), n^{0}1 , 1-23.

[5] *N.Aïssaoui et A. Benkirane,* Potentiel non linéaire dans les espaces d'Orlicz, Ann. Sci. Math. Québec 18 (1994), n^{0}2,

[6] *A. Benkirane,* Approximations de type Hedberg dans les espaces $W^m L \log L(\Omega)$ et applica tions, Annales de la faculté des sciences de Toulouse, Vol. XI n° 2 (1990), 67-78.

[7] *A. Benkirane,* Inégalités d'interpolations dans les espaces de Sobolev-Orlicz, Bull. Soc. Math. Belg. 42 (1990), 3, ser.B,285-294.

[8] *A. Benkirane and A. Elmahi,* Almost everywhere convergence of the gradients of solutions to elliptic equations in Orlicz spaces and application, to appear in Non Linear Anal. T.M.A..

[9] *A. Benkirane and A. Elmahi,*An existence theorem for a strongly nonlinear elliptic problem in Orlicz spaces, submitted.

[10] *A. Benkirane and J.-P. Gossez,* An approximation theorem for higher order Orlicz-Sobolev spaces, Studia Math.,92 (1989),pp.231-255.

[11] *L. Boccardo, D. Giachetti and F. Murat,* A generalisation of a theorem of Brézis-Browder and applications to some unilateral problems. Ann. Inst. Poincaré, Analyse non linéaire, 367-384 (1990).

[12] *H. Brézis and F. Browder,* Some properties of higher-order Sobolev spaces, J. Math. Pures Appl. (1982), 245-259.

[13] *T. Donaldson and N. Trudinger* , Orlicz-Sobolev spaces and imbedding theorems, J.Funct. Anal., 8 (1971) 367-384.

[14] *J.-P. Gossez,* Nonlinear elliptic boundary value problems for equations with rapidly(or slowly) increasing coefficients,Trans. Amer. Math. Soc. 190 (1974),pp.163-205.

[15] *J.-P. Gossez,* Some approximation properties in Orlicz-Sobolev spaces, Studia Math., 74(1982),pp.17-24.

[16] *J.-P. Gossez and V. Mustonen,* Variational inequality in Orlicz-Sobolev spaces, Nonlinear Anal.Theory Appli. 11 (1987) ,pp. 379-392.

[17] *M. Krasnosel'skii and Ya. Rutickii,* convex functions and Orlicz spaces,Noordhoff Groningen, 1969.

[18] *A. Kufner, O. John and S. Fucik,* Function spaces, Academia, Prague, 1977.

Département de Mathématiques - Faculté des sciences Dhar Mahraz - B.P. 1796 Atlas, Fès - Maroc

K BENMLIH
A result of maximum point localization for a semi-linear elliptic pde

Introduction

Let Ω be a bounded domain in $\mathbb{R}^N$. We consider the following problem :

$$(\mathcal{P}) \begin{cases} -\Delta u = f(u) & \text{in } \Omega \\ u > 0 & \text{in } \Omega \\ u = 0 & \text{on } \partial\Omega \end{cases}$$

where f is lipschitz continuous function on $\mathbb{R}^+$. Several qualitative properties of $(\mathcal{P})$, namely monotonicity, symmetry, ... have been etablished by many authors, essentially by J. Serrin [S], Gidas, Ni and Nirenberg [GNN], Berestychi and Nirenberg [BN] etc ... Here is a typical (and recent) result of [BN] :

THEOREM 1. — *Let Ω be a bounded domain in $\mathbb{R}^N$ such that :*
($\mathcal{S}$) Ω is symmetric with respect to the plane $T_0 := \{x_1 = 0\}$,
($\mathcal{C}$) Ω is convex in the x_1 direction.
Assume that f is a locally lipschitz continuous function and $u \in W^{2,N}_{loc}(\Omega) \cap C^0(\bar{\Omega})$ is a solution of $(\mathcal{P})$. Then $\frac{\partial u}{\partial x_1} > 0$ for $x_1 < 0$ and u is symmetric with respect to x_1.
In particular u attains its maximum on T_0.

The proof of this Theorem essentially consists of two key ingredients : the maximum principle and the well-known method of moving plane.

The purpose of this paper is to study the necessity of the condition $(\mathcal{C})$. For this, we consider a certain polynomial function f and construct a particular bounded open connected domain symmetric with respect to the plane T_0 but not convex in the normal direction of it (x_1). Under these considerations, we show that the solution of $(\mathcal{P})$ is not monotone in the x_1 direction and attains its maximum outside the plane of symmetry T_0.

Let $0 < r < R$ and $a > r+R$. Our idea is to consider three balls $B(-a, R)$, $B(0,r)$ and $B(a, R)$ connected by two cylindrical channels of radius ε, sufficiently small $(0 < \varepsilon << r)$ such that the domain Ω_ε so obtained is *smooth*, symmetric with respect to both the plane $\{x_1 = 0\}$ and the x_1-axis, but not convex in the x_1 direction.

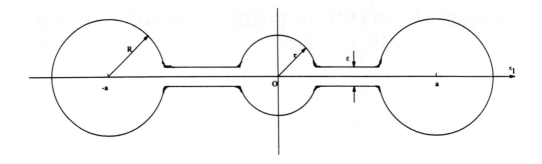

$(fig. 1)$

Throughtout this paper $\lambda_{1,\varrho}$ denotes the first strictly positive eigenvalue of the operator $-\Delta$ on $H_0^1(B(0,\varrho))$.

Our main result is the following :

THEOREM 2 . — *Let Ω_ε the domain described above $(fig.1)$ and $p > 1$. For $\lambda > \lambda_{1,r}$, let u_ε be the solution belonging to $C^{2,\theta}(\overline{\Omega}_\varepsilon)$, $\theta \in]0,1[$ of*

$$(\mathcal{P}_\varepsilon) \begin{cases} -\Delta u_\varepsilon + u_\varepsilon^p = \lambda u_\varepsilon & \text{in } \Omega_\varepsilon \\ u_\varepsilon > 0 & \text{in } \Omega_\varepsilon \\ u_\varepsilon = 0 & \text{on } \partial\Omega_\varepsilon. \end{cases}$$

Then there exists $\varepsilon_0 > 0$, sufficiently small, such that for all $0 < \varepsilon < \varepsilon_0$, the maximum of u_ε is attained outside the ball $B(0,r)$.

Let us remark, first, that by Theorem1, the solution u_ε of $(\mathcal{P}_\varepsilon)$ attains maximum on the axis (Ox_1) and more precisely, on the interval $[-a,a]$ $(cf$ Remark1$)$.

If the domain Ω_ε had been, in addition, convex in the x_1 direction, u_ε would have attained its maximum, in principle, at the intersection of several planes of symmetry of Ω_ε, i.e. at the origin. However, Theorem2 shows that the maximum point is far enough to the origin. This proves that, in the case for problem $(\mathcal{P}_\varepsilon)$, the properties of monotonocity and maximum point localization failed because of the not convexity of Ω_ε in the x_1 direction.

Our paper is organised as follows :

At first, we recall briefly the moving plane method and show how it's used to etablish some qualitative properties as in Theorem1. Next, we present a short study of existence and uniqueness of solution of $(\mathcal{P}_\varepsilon)$. In section 2, we give a convergence result of u_ε in $C(\overline{\Omega})$ to u wich is solution of a similar problem, in each of the three balls. In section 3, we show that the maximum of u on $\Omega := B(-a, R) \cup B(0, r) \cup B(a, R)$ is attained at both the points $-a$ and a and that $u(a) > u(0)$; These informations anable us to prove our main result.

1. Preliminary results

1.1 Moving plane method : description and application.

Let Ω be a bounded domain in $\mathbb{R}^N$. We suppose the plane $T_\lambda := \{(x_1, y) \in \mathbb{R} \times \mathbb{R}^{N-1} \mid x_1 = \lambda\}$ to be continuously moved to itself (in the x_1 direction) from $-\infty$ to Ω. When T_λ intersect Ω, we consider the open sets $\Sigma(\lambda) := \{(x_1, y) \in \Omega \mid x_1 < \lambda\}$ and its reflection in T_λ denoted $\Sigma'(\lambda)$.

Let u be a given function definite on Ω. To prove the monotonicity of u in the x_1 direction, we define on a new function, when $\Sigma'(\lambda)$ is contained in Ω, as :

$$v_\lambda(x_1, y) := u(2\lambda - x_1, y) \quad \text{for all } (x_1, y) \in \Sigma(\lambda)$$

Thus we consider

$$w(x, \lambda) := v_\lambda(x) - u(x) \quad \text{for } x \in \Sigma(\lambda)$$

and if we let $-a := \inf\{x_1 \mid (x_1, y) \in \Omega\}$, we wish to get

$$w(\cdot, \lambda) > 0 \quad \text{in } \Sigma(\lambda) \tag{$\star$}$$

for all $\lambda \in]-a, 0[$.

Now, if u is a solution of $(\mathcal{P})$, under the assumptions of Theorem 1, v_λ satisfies a similar equation for every $\lambda \in]-a, 0[$ and consequently, since f is lipschitz continuous on $[0, \|u\|_{L^\infty(\overline{\Omega})}]$, w_λ satisfies

$$-\Delta w(\cdot, \lambda) + c(x, \lambda)w(\cdot, \lambda) = 0 \quad \text{in } \Sigma(\lambda)$$

where $c(x, \lambda) := \dfrac{f(v_\lambda(x)) - f(u(x))}{v_\lambda(x) - u(x)} \in L^\infty(\Sigma(\lambda))$.

Using an improved form of the maximum principle in "narrow domains", Berestycki and Nirenberg have proved $(\star)$ and consequently Theorem 1 (see [BN]).

Remark 1. — The proof of Theorem 1 (given in [BN]) shows, in fact, that u increases with respect to x_1 in every $\Sigma(\lambda_*)$ such that

$$\Sigma'(\lambda) \subset \Omega \qquad \forall \lambda \in]-a, \lambda_*[.$$

So, if Ω is convex in the x_1 direction then $\lambda_* = 0$.

1.2 Study of the elliptic problem $(\mathcal{P}_\varepsilon)$

PROPOSITION 1 . — *For $\varepsilon > 0$ small enough, $(\mathcal{P}_\varepsilon)$ has a unique solution $u_\varepsilon \in C^{2,\theta}(\overline{\Omega}_\varepsilon)$, $\theta \in]0,1[$ with the uniform estimate :*

$$\|u_\varepsilon\|_{L^\infty(\Omega_\varepsilon)} \leq \lambda^{\frac{1}{p-1}} . \tag{1}$$

The idea of proof is as follows :

• Existence : the solution u_ε is obtained as a critical point of the energy associatied with the equation of $(\mathcal{P}_\varepsilon)$, on the space $X := H_0^1(\Omega_\varepsilon) \cap L^{p+1}(\Omega_\varepsilon)$, namely

$$E_\varepsilon(v) := \frac{1}{2}\int_{\Omega_\varepsilon} |\nabla v|^2 + \frac{1}{p+1}\int_{\Omega_\varepsilon} |v|^{p+1} - \frac{\lambda}{2}\int_{\Omega_\varepsilon} v^2.$$

The norm on the space X is given by $\|v\|_X := \|\nabla v\|_{L^2} + \|v\|_{L^{p+1}}$.

The solution u_ε can be assumed to be positive since $J(|u_\varepsilon|) = J(u_\varepsilon)$. Moreover, u_ε satisfies the equation of $(\mathcal{P}_\varepsilon)$ and since $\lambda > \lambda_{1,r}$ we have $u_\varepsilon \not\equiv 0$. The strong maximum principle implies that u_ε doesn't vanish in Ω_ε.

• Regularity : by a standard bootstrap argument and Schauder regularity theorem we deduce that $u_\varepsilon \in C^{2,\theta}(\overline{\Omega}_\varepsilon)$.
Now, if $x_0^\varepsilon \in \Omega_\varepsilon$ is such that $u_\varepsilon(x_0^\varepsilon) = \|u_\varepsilon\|_{L^\infty(\Omega_\varepsilon)}$ then $-\Delta u_\varepsilon(x_0^\varepsilon) \geq 0$. Therefore, with the equation of $(\mathcal{P}_\varepsilon)$ one has $u_\varepsilon^p(x_0^\varepsilon) \leq \lambda\, u_\varepsilon(x_0^\varepsilon)$ and consequently

$$u_\varepsilon(x_0^\varepsilon) \leq \lambda^{\frac{1}{p-1}}.$$

• Unicity : firstly, we give a Lemma wich etablishs some properties of the principal eigenvalue of the elliptic operator $(-\Delta + \varrho I)$, where ϱ is a positive function in L^∞.

LEMMA 1. — *Let ω be a bounded domain in $\mathbb{R}^N$, $N \geq 2$, assumed to be of class C^2. For $\varrho \in L^\infty_+(\omega)$ we denote :*

$$\lambda_1(\varrho) := \inf \left\{ \int_\omega |\nabla \psi|^2 + \int_\omega \varrho \psi^2 \mid \psi \in H^1_0(\omega), \int |\psi|^2 = 1 \right\}.$$

Then :

i) $\lambda_1(\varrho)$ is achieved at $\varphi_1 \in C^{1,\theta}(\overline{\omega})$, $\theta \in]0,1[$, wich satisfies :

$$\begin{cases} -\Delta \varphi_1 + \varrho \varphi_1 = \lambda_1(\varrho) \varphi_1 & \text{in } \Omega_\varepsilon \\ \varphi_1 > 0 & \text{in } \Omega_\varepsilon \\ \varphi_1 = 0 & \text{on } \partial \Omega_\varepsilon \end{cases}$$

ii) $\lambda_1(\varrho)$ is the unique eigenvalue of the operator $(-\Delta + \varrho I)$ of wich the eigenfunctions have a definite sign.

iii) if ϱ_1, $\varrho_2 \in L^\infty_+(\omega)$ such that $\varrho_1 \leq \varrho_2$ and $\varrho_1 \not\equiv \varrho_2$ then $\lambda_1(\varrho_1) < \lambda_1(\varrho_2)$.

For a detailed proof of this lemma, one may see [Be] for example.

Using the above Lemma, we prove easily the unicity of solution of $(\mathcal{P}_\varepsilon)$. Indeed, let u_ε and v_ε be two solutions of $(\mathcal{P}_\varepsilon)$. For $\varepsilon > 0$ arbitrarily fixed, we denote $w := u_\varepsilon - v_\varepsilon$.

If $w \not\equiv 0$, then λ is an eigenvalue of the operator $(-\Delta + \varrho_1 I)$ with

$$0 < \varrho_1 := \frac{u_\varepsilon^p - v_\varepsilon^p}{u_\varepsilon - v_\varepsilon} \leq p\lambda.$$

Consider $\lambda_1(\varrho_1)$ the first strictly positive eigenvalue of $(-\Delta + \varrho_1 I)$. So, we have :

$$\lambda_1(\varrho_1) \leq \lambda.$$

Moreover, if we denote $\varrho_0 := u_\varepsilon^{p-1}$, the problem $(\mathcal{P}_\varepsilon)$ can be written as :

$$\begin{cases} -\Delta u_\varepsilon + \varrho_0 u_\varepsilon = \lambda u_\varepsilon & \text{in } \Omega_\varepsilon \\ u_\varepsilon > 0 & \text{in } \Omega_\varepsilon \\ u_\varepsilon = 0. & \text{on} \partial \Omega_\varepsilon \end{cases}$$

Since $u_\varepsilon > 0$, the second point of Lemma 1 shows that :

$$\lambda = \lambda_1(\varrho_0). \tag{2}$$

But $\varrho_0 \leq \varrho_1$ and $\varrho_0 \neq \varrho_1$, thus the point $iii)$ of Lemma 1 implies :

$$\lambda_1(\varrho_0) < \lambda_1(\varrho_1) \leq \lambda$$

and this contradicts (2). Consequently, w must vanish everywhere in Ω_ε. ∎

Remark 2. —

$i)$ The unicity of solutions of $(\mathcal{P}_\varepsilon)$ is true for any bounded open connected domain in $\mathbb{R}^N$ with smooth boundary.

$ii)$ if we denote $v_\varepsilon(x) := u_\varepsilon(-x_1, y)$, it's clear that v_ε is also solution of $(\mathcal{P}_\varepsilon)$. The unicity of solution shows that $v_\varepsilon \equiv u_\varepsilon$, i.e. u_ε is symmetric with respect to x_1.

2. Convergence result.

Henceforth, Ω is the union of the three balls $B(-a, R)$, $B(0, R)$ and $B(a, R)$, $0 < r < R$ and we denote $\Omega := \bigcup_{i=1,2,3} B_i$. Define :

$$E(v) := \frac{1}{2} \int_\Omega |\nabla v|^2 + \frac{1}{p+1} \int_\Omega v^{p+1} - \frac{\lambda}{2} \int_\Omega v^2$$
$$= \sum_{i=1,2,3} E_i(v)$$

where $E_i(v) := \frac{1}{2} \int_{B_i} |\nabla v|^2 + \frac{1}{p+1} \int_{B_i} v^{p+1} - \frac{\lambda}{2} \int_{B_i} v^2$. Then we have the following result :

PROPOSITION 2. — *Under the assumptions of Theorem 2, there exists a function $u \in C^\infty(\Omega) \cap C^0(\overline{\Omega})$ solution of :*

$$(\mathcal{P}) \begin{cases} -\Delta u + u^p = \lambda u & \text{in } B_i \\ u > 0 & \text{in } D_i \\ u = 0 & \text{on } \partial B_i, \end{cases}$$

wich minimizes E on $H_0^1(B_i)$ for all $i = 1, 2, 3$ and we have

$$u_\varepsilon \xrightarrow{\varepsilon \downarrow 0} u \quad \text{in } C(\overline{\Omega}).$$

Proof: it's divided into several steps.

Firstly, multiplying the equation of $(\mathcal{P}_\varepsilon)$ by u_ε and using the uniform estimate (1), we deduce easily that $(u_\varepsilon)_\varepsilon$ is bounded in $H_0^1(\Omega_\varepsilon)$ and in particular in $H^1(\Omega)$. There is then a subsequence denoted again by u_ε and a function u such that:

$$u_\varepsilon \longrightarrow u \quad H^1(\Omega) \text{ weakly},$$
$$u_\varepsilon \longrightarrow u \quad L^2(\Omega),$$
$$u_\varepsilon \longrightarrow u \quad \text{a.e. in } \Omega.$$

In particular $u \geq 0$ a.e. in Ω. Moreover, using (1) and the dominated convergence theorem, we show easily that u_ε converges to u in $L^q(\Omega)$ for all $q \in [0, +\infty[$ and that:

$$-\Delta u = \lambda u - u^{p+1} \quad \text{in } \mathcal{D}'(\Omega).$$

Consequently

$$u_\varepsilon \longrightarrow u \quad \text{in } H^1(\Omega).$$

- To prove that $u \in H_0^1(\Omega)$, it suffices to show that (*cf* [Br, theorem IX.18]) there is a constant $c > 0$ such that for all $\varphi \in C_c^1(\mathbb{R}^N)$:

$$\left| \int_\Omega u\, \partial_i \varphi \right| \leq c \|\varphi\|_{L^2} \quad \forall\, i = 1, \ldots, N.$$

Indeed, consider $\varphi \in C_c^1(\mathbb{R}^N)$ (arbitrary). It follows from the Green's formula that:

$$\int_\Omega u_\varepsilon\, \partial_i \varphi = \int_{\Gamma_\varepsilon} u_\varepsilon(\sigma)\, \varphi(\sigma)\, n_{i,\varepsilon}(\sigma)\, d\sigma - \int_\Omega \partial_i u_\varepsilon(x)\, \varphi(x)\, dx$$

where $\Gamma_\varepsilon := \partial\Omega \setminus \partial\Omega_\varepsilon$. So:

$$\left| \int_\Omega u_\varepsilon\, \partial_i \varphi \right| \leq \|u_\varepsilon\|_{L^\infty} \|\varphi\|_{L^\infty} |\Gamma_\varepsilon| + \|\nabla u_\varepsilon\|_{L^2} \|\varphi\|_{L^2}$$
$$\leq c\,(|\Gamma_\varepsilon| + \|\varphi\|_{L^2})$$

Passing to the limit as $\varepsilon \to 0$ and since $|\Gamma_\varepsilon| \longrightarrow 0$, one has:

$$\left| \int_\Omega u\, \partial_i \varphi \right| \leq c \|\varphi\|_{L^2}.$$

Consequently, $u \in H_0^1(\Omega)$.

- On the other hand, by standard bootstrap argument and Schauder regularity theorem, we deduce that $u \in C^\infty(\Omega) \cap C^0(\overline{\Omega})$.
- We begin by verifying that the minimum of E_i on $H_0^1(B_i)$ is achieved at a function $u_{*,i}$, solution of

$$\begin{cases} u_{*,i} \in C^\infty(B_i) \cap C^0(\overline{B}_i) \\ -\Delta u_{*,i} + u_{*,i}^p = \lambda u_{*,i}, \quad u_{*,i} \geq 0 & \text{in } B_i \\ u_{*,i} = 0 & \text{on } \partial B_i. \end{cases}$$

Since $\lambda > \lambda_{1,r}$, we have $E_i(u_{*,i}) < 0$ and therefore $u_{*,i} \not\equiv 0$ on B_i. The strong maximum principle implies that, in fact, $u_{*,i} > 0$ on B_i.

Next, Let u_* be defined by $u_* := u_{*,i}$ on each B_i. To prove that u minimizes E_i on $H_0^1(B_i)$ for every i, it suffices to show that $u \equiv u_*$.
Set $\tilde{u}_* := u_*$ on Ω and $\tilde{u}_* := 0$ on $\Omega_\varepsilon \setminus \Omega$. We have :

$$E_\varepsilon(u_\varepsilon) \leq E_\varepsilon(\tilde{u}_*) = E(u_*).$$

If $0 < \varepsilon_1 < \varepsilon_2$, a function $\psi \in H_0^1(\Omega_{\varepsilon_1})$ may be considered as element of $H_0^1(\Omega_{\varepsilon_2})$ by setting $\psi = 0$ on $\Omega_{\varepsilon_2} \setminus \Omega_{\varepsilon_1}$. Therefore

$$\min_{v \in H_0^1(\Omega_{\varepsilon_2})} E_{\varepsilon_2}(v) < \min_{v \in H_0^1(\Omega_{\varepsilon_1})} E_{\varepsilon_1}(v),$$

on other words :

$$E_{\varepsilon_2}(u_{\varepsilon_2}) < E_{\varepsilon_1}(u_{\varepsilon_1}).$$

Thus $(E_\varepsilon(u_\varepsilon))_\varepsilon$ is a nondecreasing sequence. Since $(u_\varepsilon)_\varepsilon$ is bounded in $H^1(\Omega_\varepsilon)$ and converges to u in both $L^{p+1}(\Omega)$ and $H^1(\Omega)$, one has $E_\varepsilon(u_\varepsilon) \longrightarrow E(u)$. It follows that :

$$E(u) \leq E(u_*). \tag{3}$$

Now, we denote by u_i^* the restriction of u at the ball B_i. Then $u_i^* \in H_0^1(B_i)$ and satisfies :

$$-\Delta u_i^* + (u_i^*)^p = \lambda u_i^* \quad \text{in } B_i.$$

To show that $u_i^* \not\equiv 0$ $\forall i$, assume that $u_1^* \equiv 0$ for example. In this case, $E_1(u_1^*) = 0$ and therefore :

$$E(u) = E_2(u_2^*) + E_3(u_3^*)$$
$$\geq E_2(u_{*,2}) + E_3(u_{*,3})$$
$$> \sum_{i=1}^{3} E_i(u_{*,i}) = E(u_*)$$

But this contradicts (3). We deduce, thereby, that $u_i^* \not\equiv 0$ and by the maximum principle (since $u \geq 0$) that $u_i^* > 0$ for all i. The unicity of solutions of $(\mathcal{P})$ (see Remark 2, i)) implies that $u_i^* = u_{*,i}$ $\forall i \in \{1,2,3\}$, i.e. $u = u_*$.

• Finaly, to show that $(u_\varepsilon)_\varepsilon$ converges uniformly to u on $\overline{\Omega}$, we need the following two lemmas :

LEMMA 2 . — *Let ω be a bounded open connected domain in $\mathbb{R}^N$, $N \geq 2$, having a smooth boundary and $\varphi, \psi \in H^1(\omega)$ such that :*

$$\begin{cases} -\Delta\varphi + \varphi^p \leq -\Delta\psi + \psi^p & \text{in } \omega \\ \varphi \leq \psi & \text{on } \partial\omega, \end{cases}$$

then $\varphi \leq \psi$ in ω.

LEMMA 3 . — *Let $\Omega_1 \subset \Omega_2$ be two bounded open connected domains in $\mathbb{R}^N$, $N \geq 2$, assumed to be of class C^2 and $u_1 \in H^1(\Omega_1)$, $u_2 \in H^1(\Omega_2)$ solutions of :*

$$(\mathcal{P}_{\Omega_k}) \begin{cases} -\Delta u_k + u_k^p = \lambda u_k & \text{in } \Omega_k \\ u_k > 0 & \text{in } \Omega_k \\ u_k = 0 & \text{on } \partial\Omega_k \end{cases}$$

for $k = 1, 2$. Set

$$z := \begin{cases} u_1 & \text{in } \Omega_1, \\ 0 & \text{in } \Omega_2 \setminus \Omega_1. \end{cases}$$

a) *z is a subsolution of the equation of $(\mathcal{P}_{\Omega_2})$ in the following sense :*

$$-\Delta z + z^p - \lambda z \leq 0 \qquad \text{in } \mathcal{D}'(\Omega_2).$$

b) Let $(v_n)_{n\in\mathbb{N}}$ and $(w_n)_{n\in\mathbb{N}}$ be two sequences in $H_0^1(\Omega_2)$ by :

$$\begin{cases} v_0 := z \\ \forall n \geq 1 \quad -\Delta v_n + v_n^p = \lambda v_{n-1}, \end{cases}$$

$$\begin{cases} w_0 := \lambda^{\frac{1}{p-1}} \mathbb{1}_{\Omega_2} \\ \forall n \geq 1 \quad -\Delta w_n + w_n^p = \lambda w_{n-1}. \end{cases}$$

Then :

i) $\forall n \geq 1 \quad z \leq v_{n-1} \leq v_n \leq w_n \leq w_{n-1} \leq \lambda^{\frac{1}{p-1}} \quad on \ \Omega_2.$

ii) *For some* $\alpha \in (0,1)$ *one has* :

$$v_n \xrightarrow{C^{1,\alpha}(\overline{\Omega}_2)} u_2$$

c) $u_1 \leq u_2 \quad in \ \Omega_1$

By using these lemmas, the uniform convergence of u_ε on $\overline{\Omega}$ to u is proved as follows :

⋆) if $0 < \varepsilon_1 < \varepsilon_2$ then $\Omega_{\varepsilon_1} \subset \Omega_{\varepsilon_2}$. By the property c) of Lemma 3, we have

$$0 \leq u_{\varepsilon_1} \leq u_{\varepsilon_2} \quad in \ \Omega_{\varepsilon_1}.$$

⋆) Since the sequence $(u_\varepsilon)_\varepsilon$ is decreasing (with respect to $\frac{1}{\varepsilon}$) and bounded below, there exists a function u_* such that for all $x \in \overline{\Omega}$:

$$u_\varepsilon(x) \xrightarrow{\varepsilon \downarrow 0} u_*(x) \geq 0.$$

This implies in particular that

$$-\Delta u_* + u_*^p = \lambda u_* \quad in \ \mathcal{D}'(\Omega).$$

On the other hand, u_ε (subsequence of u_ε) converges in $L^2(\Omega)$ to u and to u_*, therefore $u = u_*$ a.e. in Ω and consequently $u_* \in H_0^1(\Omega)$.
By regularity theorem, as before, we conclude that $u_* \in C(\overline{\Omega})$.

Finally, we use the Dini's theorem : since $(u_\varepsilon)_\varepsilon$ is monotone and converges to u_* at each point and since $\overline{\Omega}$ is a compact set in R^N, the sequence $(u_\varepsilon)_\varepsilon$ converges uniformly to u on $\overline{\Omega}$. In other words (recall that $u = u_*$) :

$$\|u_\varepsilon - u\|_{C(\overline{\Omega})} \xrightarrow{\varepsilon \downarrow 0} 0.$$

This completes the proof of Proposition 2. ∎

Proof of Lemma 2 :

We have

$$\begin{cases} -\Delta(\varphi - \psi) + (\varphi^p - \psi^p) \leq 0 & \text{in } \omega \\ (\varphi - \psi) \leq 0 & \text{on } \partial\omega. \end{cases}$$

Multiplying the first inequality by $(\varphi - \psi)^+$ and integrating, we obtain :

$$\int_\omega \nabla(\varphi - \psi).\nabla(\varphi - \psi)^+ - \int_{\partial\omega} \frac{\partial(\varphi - \psi)}{\partial n}(\varphi - \psi)^+(\sigma)\, d\sigma$$
$$+ \int_\omega (\varphi^p - \psi^p)(\varphi - \psi)^+ dx \leq 0,$$

it follows that :

$$\int_\omega |\nabla(\varphi - \psi)^+|^2 + \int_\omega (\varphi^p - \psi^p)^+(\varphi - \psi)^+ dx \leq 0.$$

Therefore $(\varphi - \psi)^+ = 0$ in ω, i.e.

$$\varphi \leq \psi \qquad \text{in } \omega. \blacksquare$$

Proof of Lemma 3 :

a) Let $\varphi \in \mathcal{D}(\Omega_2)$, $\varphi \geq 0$ then

$$<-\Delta z, \varphi> = -<z, \Delta\varphi>$$
$$= \int_{\Omega_2} \nabla z.\nabla\varphi - \int_{\partial\Omega_2} \frac{\partial\varphi}{\partial n}(\sigma) z(\sigma)\, d\sigma$$
$$= \int_{\Omega_1} \nabla u_1.\nabla\varphi$$
$$= -\int_{\Omega_1} \Delta u_1\, \varphi + \int_{\partial\Omega_1} \frac{\partial u_1}{\partial n}(\sigma) \varphi(\sigma)\, d\sigma.$$

In view of the Hopf lemma ([GT], Lemma 3.4) we have

$$\forall \sigma \in \partial\Omega_1 \qquad \frac{\partial u_1}{\partial n}(\sigma) < 0.$$

Since, in addition, u_1 satisfies the equation of (P_{Ω_1}) one has

$$< -\Delta z, \varphi > \leq \lambda \int_{\Omega_1} u_1 \varphi - \int_{\Omega_1} u_1^{p+1} \varphi$$
$$= \lambda \int_{\Omega_2} z \varphi - \int_{\Omega_2} z^{p+1} \varphi.$$

Thus

$$< -\Delta z + z^p - \lambda z, \varphi > \leq 0 \qquad \forall \varphi \in \mathcal{D}(\Omega_2), \ \varphi \geq 0.$$

b-i) Using the first point **a)** we prove easily, by a simple recurrence argument, that $(v_n)_n$ and $(w_n)_n$ are nondecreasing and decreasing sequences respectively. Moreover, since $\|z\|_{L^\infty(\Omega_2)} \leq \lambda^{\frac{1}{p-1}}$, it is clear that $v_n \leq w_n$ in Ω_2 for all $n \in \mathbb{N}$.

b-ii) By **b-i)** we have, in particular, the uniform estimate

$$\|v_n\|_{L^q(\Omega_2)} \leq k_0 \qquad \forall q \in [1,\infty]$$

where $k_0 = k_0(\lambda, p, |\Omega_2|)$. Since $v_n = 0$ on $\partial\Omega_2$, one has for all $n \in \mathbb{N}^*$:

$$\|v_n\|_{W^{2,q}(\Omega_2)} \leq k_1 ;$$

Choosing q large enough, the Morrey-Sobolev imbedding implies that :

$$\|v_n\|_{C^{1,\beta}(\overline{\Omega}_2)} \leq k_2 .$$

where $\beta \in]0,1[$, k_1 and k_2 are two constants independents of n.

Let $0 < \alpha < \beta$, by compactness of the imbedding $C^{1,\beta}(\overline{\Omega}_2)$ into $C^{1,\alpha}(\overline{\Omega}_2)$ there exists $v_* \in C^{1,\alpha}(\overline{\Omega}_2)$ and a subsequence $(v_{n_i})_i$ of $(v_n)_n$ such that :

$$v_{n_i} \xrightarrow{C^{1,\alpha}(\overline{\Omega}_2)} v_*.$$

Therefore, v_* is a solution of (P_{Ω_2}) and the unicity result (Proposition 1) implies that $v_* = u_2$ and that, in fact, the whole sequenece $(v_n)_n$ converges in $C^{1,\alpha}(\overline{\Omega_2})$ to u_2.

c) It's clear that b-i) and b-ii) give $z \leq u_2$ in Ω_2 i.e.

$$u_1 \leq u_2 \quad \text{in } \Omega_2. \blacksquare$$

3 - Behaviour of $\|u\|_{L^\infty(B(0,r))}$ with respect to r.

Our choice of the geometry of Ω_ε (the hypothesis $r < R$ precisely) is based on the fact that the maximum value of the solution u of $(\mathcal{P})$ on a ball $B(0,r)$ is strictly increasing with respect to the radius r. Indeed

PROPOSITION 3. — For $r > 0$, we consider $B_r := B(0,r) \subset \mathbb{R}^N, N \geq 2$ and $v_r \in H_0^1(B_r)$ the solution of the problem

$$(P_r) \begin{cases} -\Delta v_r + v_r^p = \lambda v_r & \text{in } B_r \\ v_r > 0 & \text{in } B_r \\ v_r(x) = 0. & \text{on } \partial B_r \end{cases}$$

Then $r \longmapsto \|v_r\|_{L^\infty(B_r)}$ is strictly increasing.

Proof: It is clear that every solution of (P_r) is radially symmetric and attains its maximum at the origin (*cf* Theorem 1, for example). Therefore, setting $v_r(x) = f(t)$ where $t := |x|$, the function f satisfies the ordinary differential equation :

$$(D_r) \begin{cases} -f'' - \dfrac{N-1}{t} f' = \lambda f - f^p & f > 0 \quad t \in\,]0, r[\\ f'(0) = 0 \\ f(0) = v_r(0) \end{cases}$$

For $R > r$, let v_R be the solution of (P_R). Then, by setting $v_R(x) = g(t)$, g is solution of (D_R). By the property c) of Lemma 3, one has $v_r \leq v_R$ on $B(0,r)$. Now, we wish to show that $v_r(0) < v_R(0)$.

Indeed, if $f(0) = g(0)$ the function g would have been a solution of (D_r). Thus, by the unicity of solution of (D_r) one has $f \equiv g$ on $[0, r]$. In particular $v_R(x) = 0$ on ∂B_r and this is obviously false. Consequently, for $0 < r < R$ we have :

$$0 < v_r(0) < v_R(0) < \lambda^{\frac{1}{p-1}}. \blacksquare$$

3 - Proof of main result.

Let u and u_ε be the solutions given by Theorem 2. In view of Proposition 3, the maximum of u on Ω is attained at two points $-a$ and a, and we have $u(-a) = u(a) > u(0)$. By the uniform convergence of u_ε to u in $\overline{\Omega}$, there is $\varepsilon_0 > 0$ sufficiently small such that :

$$\forall\, 0 < \varepsilon < \varepsilon_0, \quad \forall\, x \in B(0, r) \quad |u_\varepsilon(x) - u(x)| < \frac{u(a) - u(0)}{2}.$$

Since $\|u\|_{L^\infty(B(0,r))} = u(0)$, this implies that

$$\forall\, x \in B(0, r) \quad u_\varepsilon(x) < \frac{u(a) + u(0)}{2}.$$

On the other hand, $u_\varepsilon(a) \longrightarrow u(a)$ as $\varepsilon \downarrow 0$ then

$$\forall\, 0 < \varepsilon < \varepsilon_0, \quad u_\varepsilon(a) > \frac{u(a) + u(0)}{2}.$$

Consequently

$$\forall\, 0 < \varepsilon < \varepsilon_0, \quad \forall\, x \in B(0, r) \quad u_\varepsilon(x) < u_\varepsilon(a).$$

It follows in particular that for each $0 < \varepsilon < \varepsilon_0$, the maximum of u_ε on Ω_ε is attained outside the ball $B(0, r)$ and this completes the proof of Theorem 2. $\blacksquare$

Acknowledgements - I would like to thank Professor O. Kavian for his valuable suggestions.

References

[Be] BENMLIH Kh., Thèse de Doctorat de l'Université Henri Poincaré, Nancy 1, France, 1994.

[BN] BERESTYCKI H. & NIRENBERG L., On the method of moving planes and the sliding method, Boll. Soc. Brasil Mat. Nova 22 (1991), 1-37.

[Br] BREZIS H., *Analyse fonctionnelle, Théorie et applications*; Edition Masson, Paris 1983.

[GNN] GIDAS B., NI W. M. & NIRENBERG L., Symmetry and related properties via the maximum principle, Comm. Math. Phys. 68 (1979), 209-243.

[GT] GILBARG B. & TRUDINGER N. S., *Elliptic partial differential equations of second order*, Springer-Verlag, Heidelberg 1983.

[S] SERRIN J., A symmerty problem in potentiel theory; Arch. Rat. Mech. & Analysis 43 (1971), 304-318.

ADDRESS

Khalid BENMLIH, Université Sidi Mohamed Ben Abdellah, Faculté des sciences juridiques économiques et sociales, Département des sciences économiques, BP 42A Dhar Mehraz, FES, MAROC.

J BERKOVITS AND V MUSTONEN
Existence results for semilinear equations with normal linear part and applications to telegraph equations

1. Introduction

In this paper we shall deal with the existence of nontrivial solutions of semilinear telegraph equations of the form

$$\begin{cases} u_{tt} - u_{xx} + \beta u_t = g(x,t,u), (x,t) \in\,]0,\pi[\times \mathrm{R} \\ u(0,t) = u(\pi,t) = 0, t \in \mathrm{R} \\ u(x,t+2\pi) = u(x,t), (x,t) \in\,]0,\pi[\times \mathrm{R} \end{cases} \quad \text{(TE)}$$

where $\beta \neq 0$ is a constant and the function $g(x,t,s)$ is 2π-periodic in t, measurable in (x,t) for each $s \in \mathrm{R}$ and continuous in s for almost all $(x,t) \in\,]0,\pi[\times \mathrm{R}$. The problem (TE) is semilinear in the sense that

$$a \leq \frac{g(x,t,s)}{s} \leq b, s \neq 0, (x,t) \in \Omega$$

for some constants a and b with $\Omega \in\,]0,\pi[\times]0,2\pi[$. We shall see first that some "interaction" between g and the linear part is necessary for the existence of nontrivial solutions for (TE). This note is devoted to the case when the function $h(\cdot,\cdot,s) = s^{-1}g(\cdot,\cdot,s)$ *crosses some eigenvalues* of the linear part as $|s|$ goes from 0 to ∞. In fact, we are interested in the existence of weak solutions of (TE) and therefore it is relevant to study more general equations of the type

$$Au = N(u) \quad \text{(E)}$$

in a closed subspace H of $L^2(\Omega)$, where Ω is a bounded domain in R^n, A is densely defined closed normal linear operator with Im $A = (\mathrm{Ker}\ A)^\perp$ having a compact partial inverse and N is a nonlinear Nemytski operator generated by some Carathéodory function g. There are two main features of the linear part which create difficulties in the study of equation (E); lack of compactness when Ker A is infinite dimensional and the fact that variational methods are not available when A is not self adjoint.

If A is self adjoint, the theory of the existence of nontrivial solutions for (E) is very rich and extensively studied in various cases (nonresonance, resonance, jumping or crossing nonlinearity; see for instance [1], [2], [6], [8], [10], [11], [13] and the references mentioned therein). The case where A is not self adjoint is less studied. A substantial contribution in this direction is due to Brezis and Nirenberg [5] where they treat a very general class of equations (E) in resonance and nonresonance cases even when Ker A is infinite dimensional. Results on the periodic solutions of semilinear telegraph equations had been obtained before [5] by Mawhin [12] and Fučik and Mawhin [7] in the nonresonance case.

We shall treat the crossing case using degree arguments introduced in our previous paper [3]. Our basic abstract existence result is obtained when A is normal, $\operatorname{Ker} A$ may be infinite dimensional and just one simple eigenvalue is crossed. For the telegraph equation (TE) we show that this result can be applied in more general situations if g meets certain symmetry conditions. Similar results for semilinear wave equations ($\beta = 0$) are shown in [3] (cf. [1], [2]). As indicated by Hetzer [9] we have, of course, to take the complex part of the spectrum of A into consideration. Note that for the telegraph equation (TE) the linear part is invertible and thus the classical Leray-Schauder degree may be applied.

2. Prerequisites

Let H be a real separable Hilbert space with inner product $\langle \cdot, \cdot \rangle$ and corresponding norm $\|\cdot\|$ and let $A : D(A) \subset H \to H$ be a densely defined closed, normal linear operator with $\operatorname{Im} A = (\operatorname{Ker} A)^\perp$. The inverse A_0^{-1} of the restriction $A_0 = A|_{D(A) \cap \operatorname{Im} A}$ is a bounded linear operator on $\operatorname{Im} A$. We shall assume that A_0^{-1} is compact. In particular, we are interested in the case where A is not self adjoint, i.e., $A^* \neq A$. Since we are interested in the interaction with the spectrum of A, we include the complex spectrum of A into consideration. We denote the complexification of H by H_C. The linear structure in H_C is given by $z + w := (u + x) + i(v + y)$ and $\lambda z := (au - bv) + i(av + bu)$ for all $z = u + iv$, $w = x + iy \in H_C$ and $\lambda = a + ib \in \mathbb{C}$. The inner product in H_C is defined by

$$\langle z, w \rangle_C = \langle u, x \rangle + \langle v, y \rangle + i(\langle x, v \rangle - \langle u, y \rangle).$$

Moreover we define the complex linear operator $A_C : D(A_C) \subset H_C \to H_C$ by setting $D(A_C) = \{u + iv \in H_C \mid u, v \in D(A)\}$ and $A_C(u + iv) = Au + iAv$ for all $u + iv \in D(A_C)$. It is clear that $\operatorname{Im} A_C = (\operatorname{Ker} A_C)^\perp$, A_C is normal and the inverse of $A_C|_{D(A_C) \cap \operatorname{Im} A_C}$ is compact. Consequently, A_C has a pure point spectrum $\sigma(A_C) = \{\mu_j\}_{j \in \Lambda}$, where Λ is some subset of $\mathbb{Z}$. Each nonzero eigenvalue μ_j has finite multiplicity $m(\mu_j)$ but the multiplicity $m(0) = \dim \operatorname{Ker} A_C$ may be infinite. Moreover, there exists an orthonormal basis $\{\phi_{jk} \mid j \in \Lambda, k = 1, 2, \ldots, m(\mu_j)\}$ of H_C such that

$$A_C \phi_{jk} = \mu_j \phi_{jk}, \qquad k = 1, \ldots, m(\mu_j), j \in \Lambda.$$

For each $z \in D(A_C)$ we have spectral representation

$$A_C z = \sum_{j \in \Lambda} \sum_{k=1}^{m(\mu_j)} \mu_j \langle z, \phi_{jk} \rangle_C \phi_{jk}.$$

Note that $A_C(D(A)) \subset H$ and hence for all $u \in D(A)$

$$\langle A(u), u \rangle = \langle A_C u, u \rangle_C = \sum_{j,k} \operatorname{Re} \mu_j |\langle u, \phi_{jk} \rangle_C|^2. \tag{2.1}$$

If τ is any real number such that $\tau \notin \sigma(A_c)$, $A - \tau I$ is invertible and we easily get

$$\|(A - \tau I)^{-1}\| = \frac{1}{\text{dist}(\tau, \sigma(A_c))}. \tag{2.2}$$

In the proof of our main result we shall assume that A_C has a simple real eigenvalue λ_0 such that $\text{Re } \mu_j \neq \lambda_0$ for all $j \in \Lambda$, $\mu_j \neq \lambda_0$. Indeed, we need the condition

($\mathcal{A}$) There exists $\lambda_0 \in \sigma(A_c) \cap \mathbb{R}$ such that $m(\lambda_0) = 1$ and $\underline{\theta} < \lambda_0 < \overline{\theta}$, where

$$\underline{\theta} = \sup_{\mu_j \neq \lambda_0} \{\text{Re } \mu_j \mid \text{Re } \mu_j \leq \lambda_0\}$$

$$\overline{\theta} = \inf_{\mu_j \neq \lambda_0} \{\text{Re } \mu_j \mid \text{Re } \mu_j \geq \lambda_0\}$$

It is obvious that we can assume that the eigenvector ϕ_0 associated to λ_0 is real in the sense that $\phi_0 \in H$. If ($\mathcal{A}$) holds we denote

$$H_1 = \overline{\text{sp}}\{\phi_{jk} \mid \text{Re } \mu_j \leq \underline{\theta}\} \cap H$$
$$H_2 = \text{sp}\{\phi_0\}$$
$$H_3 = \overline{\text{sp}}\{\phi_{jk} \mid \text{Re } \mu_j \geq \overline{\theta}\} \cap H.$$

The orthogonal projection from H to H_i is denoted by P_i ($i = 1, 2, 3$). Note that the subspaces H_i are mutually orthogonal and $H_1 \oplus H_2 \oplus H_3 = H$. By (2.1) we have the following crucial estimates

$$\langle Au, u \rangle \leq \underline{\theta}\|u\|^2 \text{ for all } u \in D(A) \cap H_1 \tag{2.3}$$

$$\langle Au, u \rangle \geq \overline{\theta}\|u\|^2 \text{ for all } u \in D(A) \cap H_3 \tag{2.4}$$

Remark. If $\text{Re } \mu > \lambda_0$ for all $\mu \in \sigma(A_c)$, $\mu \neq \lambda_0$, then $H_1 = \{0\}$. Similarly, if for all $\mu \in \sigma(A_c)$, $\mu \neq \lambda_0$, $\text{Re } \mu < \lambda_0$, then $H_3 = \{0\}$.

Next we introduce the nonlinear part of equation (E). To this end we assume that $\Omega \subset \mathbb{R}^n$ is a bounded domain and H is a closed subspace of the real Hilbert space $L^2(\Omega)$. Let g be a Carathéodory function $(x, s) \to g(x, s)$ from $\Omega \times \mathbb{R}$ to $\mathbb{R}$, i.e. measurable in x for all s and continuous in s for almost all $x \in \Omega$. We assume that there exist constants a and b such that

$$a \leq \frac{g(x, s)}{s} \leq b, \text{ a.e. in } \Omega, s \neq 0 \tag{g_1}$$

Then g generates a Nemytski operator $N : H \to L^2(\Omega)$ by $N(u) = g(\cdot, u)$, $u \in H$ and N is continuous and bounded in the sense that N takes bounded sets into bounded sets. Moreover, if $g(x, \cdot)$ is nondecreasing, N is monotone, i.e.,

$$\langle N(u) - N(v), u - v \rangle \geq 0 \text{ for all } u, v \in H.$$

In the sequel we assume that H is invariant under N, i.e., $N(H) \subset H$. If $H \neq L^2(\Omega)$ this assumption usually means some restrictions for g as we indicate in section 5. We are interested in the existence of nontrivial solutions of the equations

$$A(u) = N(u), u \in D(A) \subset H. \tag{E}$$

In [3] we introduced a degree theory for a class of mappings relevant to the study of equation (E). In fact, if Ker A is finite dimensional, then the classical Leray-Schauder degree can be applied to (E) for any N satisfying (g_1). In the case A has infinite dimensional kernel we need the monotonicity condition for N.

3. Preliminary results

In this section we assume that Ω is a bounded domain in $\mathbb{R}^n$, H is a closed subspace of $L^2(\Omega)$ and $A: D(A) \subset H \to H$ is a densely defined closed, normal linear operator as introduced in section 2. We also assume that the mapping $N: H \to H$ is generated by a Carathéodory function g satisfying (g_1). The following result shows that some interaction between A and N is necessary for the existence of nontrivial solutions for (E).

Theorem 1. Let g satisfy the condition (g_1) with $[a,b] \cap \sigma(A) = \emptyset$ and assume there exists $\tau \in]a,b[$ such that

$$\max\{\tau - a, b - \tau\} < \text{dist}(\tau, \sigma(A_c)). \tag{3.1}$$

Then the equation $Au = N(u)$ admits only the trivial solution $u = 0$.

Proof. If $Au = N(u)$ for some $u \in D(A)$, then $u = (A - \tau I)^{-1}(N(u) - \tau u)$. By (2.2) and (g_1) we have

$$\|u\| \leq \frac{\|N(u) - \tau u\|}{\text{dist}(\tau, \sigma(A_c))} \leq \frac{\eta_\tau}{\text{dist}(\tau, \sigma_c(A))} \|u\|,$$

where $\eta_\tau = \max\{\tau - a, b - \tau\}$. In view of (3.1) $u = 0$.

Remark. If A is self adjoint then $[a,b] \cap \sigma(A) \neq \emptyset$ is a necessary condition for the existence of nontrivial solutions.

It is interesting to notice that if dim Ker $A < \infty$, then the conditions of Theorem 1 imply that $L - N$ is onto, i.e., the equation

$$Au = N(u) + h, \quad u \in D(A) \subset H \tag{E}_h$$

admits a solution for all $h \in H$. Analogously, if dim Ker $A = \infty$ the same conclusion holds provided $(\text{sgn } b)g(x, \cdot)$ is nondecreasing. Indeed, using the estimates of the proof of Theorem 1 we easily see that there exists $R > 0$ such that

$$Au - (1-t)\tau u - tN(u) \neq th$$

for all $0 \leq t \leq 1$, $u \in D(A)$, $\|u\| = R$. The assertion follows by the homotopy arguments of [3].

We give some equations where the linear part satisfies the conditions imposed above on A and A is not self adjoint.

(a) Telegraph equation in $\mathbb{R}^m$ ($m \geq 1$)

$$\begin{cases} u_{tt} - \Delta u + \beta u_t = g(x,t,u), (x,t) \in]0,\pi[^m \times \mathbb{R} \\ u(x,t) \text{ is } 2\pi\text{-periodic in } t \text{ with zero} \\ \text{boundary condition in } x \end{cases} \quad (3.2)$$

For $m = 1$ we have the equation (TE). The boundary condition can be replaced by 2π-periodicity condition (double-periodic case).

(b) Beam equation with linear damping

$$\begin{cases} u_{tt} + u_{xxxx} + \beta u_t = g(x,t,u), (x,t) \in]0,\pi[\times \mathbb{R} \\ u(x,t) \text{ is } 2\pi\text{-periodic in } t \text{ with zero} \\ \text{boundary condition in } x \end{cases} \quad (3.3)$$

(c) Vibrating string equation with nonconstant density

$$\begin{cases} u_{tt} - \frac{\partial}{\partial x}[d(x)u_x] + \beta u_t = g(x,t,u), (x,t) \in]0,\pi[\times \mathbb{R} \\ u(x,t) \text{ is } 2\pi \text{ periodic in } t \text{ with zero} \\ \text{boundary condition in } x \end{cases} \quad (3.4)$$

where $d_1 \leq d(x) \leq d_2$ for all $x \in [0,\pi]$ with some positive constants d_1 and d_2.

More generally, let $G \subset \mathbb{R}^m$ be a bounded smooth domain and $X = L^2(G)$. Assume that $L : D(L) \subset X \to X$ is a densely defined (differential) operator, which is closed self-adjoint and Im $L = (\text{Ker } L)^{\perp}$ such that the inverse of the restriction of L to Im L is compact. Then $\sigma(L) = \{\lambda_j \mid j \in I\}$ with $I \subset \mathbb{Z}$ and each $\lambda_j \neq 0$ has finite multiplicity $m(\lambda_j)$ but Ker L may be infinite dimensional. The eigenfunctions $\{w_{jl} \mid j \in I, l = 1, \ldots, m(\lambda_j)\}$ form an orthonormal basis in X. Denoting $\Omega = G \times]0, 2\pi[$ and $H = L^2(\Omega)$ it is clear that the set

$$\{w_{jl} \sin kt \mid k \in \mathbb{Z}_+\} \cup \{w_{jl} \cos kt \mid k \in \mathbb{N}\}$$

generates an orthonormal basis $\{\phi_{jlk}\}$ in H. Then the generalized telegraph operator $u \to u_{tt} + Lu + \beta u_t$ has in H an abstract realization

$$Au = \sum_{j,l,k}\{(\lambda_j - k^2)u_{jlk} - \beta k u_{j,l,-k}\}\phi_{jlk},$$

where $u_{jlk} = \langle u, \phi_{jlk}\rangle_H$. A is densely defined in H and has the complex spectrum

$$\sigma(A_c) = \{\mu_{jk} = (\lambda_j - k^2) + i\beta k \mid j \in I, k \in \mathbb{Z}\}$$

Note that the spatial part L is not necessarily elliptic. Note also that the condition ($\mathcal{A}$) means in the above setting that $\lambda_0 \in \sigma(A_c) \cap \mathbb{R} = \sigma(A)$ with $m(\lambda_0) = 1$ and $\operatorname{Re} \mu_{jk} = \lambda_j - k^2 \neq \lambda_0$ for all $j \in I$, $k \neq 0$. For instance, the equation

$$u_{tt} - (u_{xx} - u_{yy}) + \beta u_t = g(x, y, t, u), (x, y, t) \in]0, \pi[^2 \times \mathbb{R}$$

with periodicity and zero-boundary conditions gives rise to the equation of the type (E). Note, however, that the condition ($\mathcal{A}$) is not satisfied.

4. Existence theorem

We shall study the existence of multiple solutions of the equation

$$Au = N(u), u \in D(A) \subset H, \tag{E}$$

where H is a closed subspace of $L^2(\Omega)$, $A : D(A) \subset H \to H$ is a densely defined normal linear operator satisfying the condition ($\mathcal{A}$) and N is a Nemytski operator generated by a Carathéodory function g which satisfies the condition (g_1). We also assume that the following conditions hold

$$N(H) \subset H, H_2 = \operatorname{sp}\{\phi_0\} \subset L^\infty(\Omega) \tag{g_2}$$

There exist constants c, d and $\bar{a}$ such that
$a < \bar{a} < \lambda_0$ and

$$a \leq \frac{g(x, s)}{s} \leq \bar{a} \text{ for all } |s| > d, \text{ a.e. } x \in \Omega, \tag{g_3}$$

$$\lambda_0 \leq \frac{g(x, s)}{s} \leq b \text{ for all } 0 < s \leq c, \text{ a.e. } x \in \Omega$$

In the case Ker A is infinite dimensional we need the following condition

$$\lambda_0 \neq 0 \text{ and } \operatorname{sgn} \lambda_0 g(x, s) \text{ is nondecreasing in } s \tag{g_4}$$

Our main result is based on the following continuation theorem which is a special case of the results in [3] (cf. [13])

Continuation theorem. Let $G \subset H$ be an open bounded convex set with $0 \notin \overline{G}$, let $A : D(A) \subset H \to H$ satisfy ($\mathcal{A}$) and let $N : H \to H$ satisfy the conditions (g_1), (g_2) and (g_3). Assume that $w \in (A - \lambda_0 I + P_2)(G \cap D(A))$ is given and

$$Au - tN(u) - (1-t)(\lambda_0 u - P_2 u) \neq (1-t)w \tag{4.1}$$

for all $u \in \partial G \cap D(A)$ and $0 < t < 1$. Then we have:
(a) If dim Ker $A < \infty$, the equation $Au = N(u)$ admits at least one (nontrivial) solution $u \in \overline{G} \cap D(A)$

(b) If dim Ker $A = \infty$ and N satisfies (g_4), the equation $Au = N(u)$ admits at least one (nontrivial) solution $u \in \overline{G} \cap D(A)$.

We shall show that under the conditions above there are distinct sets G_+ and G_- such that for $w = \phi_0 \in G_+$ and $w = -\phi_0 \in G_-$, respectively, the condition (4.1) holds. Recall that $\phi_0 \in H_2$ is the eigenfunction associated to the crossed eigenvalue λ_0. Indeed, we denote

$$F_t := A - tN - (1-t)(\lambda_0 I - P_2)$$

and

$$S := \{u \in D(A) \mid F_t(u) = (1-t)w \text{ for some } 0 \leq t \leq 1\}.$$

Note that $A - \lambda_0 I + P_2$ is one to one and $u = \pm\phi_0$ is the only solution to the equation $Au - \lambda_0 u + P_2 u = \pm\phi_0$. Note also that $0 \notin S$ but $w = \pm\phi_0 \in S$. Using (2.3), (2.4), (g_1) and (g_3) we get for all $u \in S$

$$\begin{aligned} t\langle N(u) - \lambda_0 u, P_1 u\rangle &\leq -(\lambda_0 - \underline{\theta})\|P_1 u\|^2 \\ t\langle N(u) - \lambda_0 u, P_2 u\rangle &= (1-t)\langle P_2 u - w, P_2 u\rangle \\ t\langle N(u) - \lambda_0 u, P_3 u\rangle &\geq (\overline{\theta} - \lambda_0)\|P_1 u\|^2 \end{aligned} \quad (4.2)$$

In order to handle the component $P_2 u$ we consider separately the cases $\langle P_2 u - w, P_2 u\rangle \geq 0$ and $\langle P_2 u - w, P_2 u\rangle \leq 0$. Denoting

$$D^+ := \{u \in H \mid \langle P_2 u - w, P_2 u\rangle \geq 0\} = \{u \in H \mid P_2 u = \mu w, \mu \leq 0 \text{ or } \mu \geq 1\}$$
$$D^- := \{u \in H \mid \langle P_2 u - w, P_2 u\rangle < 0\} = \{u \in H \mid P_2 u = \mu w, 0 < \mu < 1\}$$

we can recover from [4] the following estimates.

Lemma 1. There exist constants $R > 1$ and $0 < \rho < 1$ such that $S \cap D^- \subset \{u \in H \mid \|u\| < R, P_2 u = \mu w, \mu > \rho\}$ and $S \cap D^+ \subset \{u \in H \mid \|u\| < R\}$.

Now we are in the position to present

Theorem 2. Let $A : D(A) \subset H \to H$ satisfy the condition $(\mathcal{A})$ and let N satisfy the conditions (g_1), (g_2) and (g_3). Then we have:
(a) If dim Ker $A < \infty$, the equation $Au = N(u)$ admits at least two nontrivial solutions $u \in D(A)$ with $P_2 u \neq 0$.
(b) If dim Ker $A = \infty$ and N satisfies (g_4), the equation $Au = N(u)$ admits at least two nontrivial solutions $u \in D(A)$ with with $P_2 u \neq 0$.

Proof. We define the sets

$$G_\pm := \{u \in H \mid \|u\| < R, P_2 u = \pm\mu\phi_0, \mu > \rho\},$$

where R and ρ are the constants obtained in Lemma 1. Obviously $\overline{G}_+ \cap \overline{G}_- = \emptyset$ and both sets are open bounded and convex. Using the estimates of Lemma 1 it is not hard to see that the continuation theorem applies and hence there exist $u_1 \in \overline{G}_+ \cap D(A)$ and $u_2 \in \overline{G}_- \cap D(A)$ which are distinct nontrivial solutions of the equation $Au = N(u)$.

Remark. If $g(x,\cdot)$ is odd, as we frequently assume in applications, it is of course possible that $u_2 = -u_1$ and Theorem 2 gives one pair $\{u, -u\}$ of nontrivial solutions.

5. Applications to one-dimensional telegraph equation

We consider now more closely the equation (TE) described in section 1. The telegraph operator $u \to u_{tt} - u_{xx} + \beta u_t$ has in $L^2(\Omega)$ abstract realization of the form

$$Au = \sum_{j=1}^{\infty} \sum_{k=-\infty}^{\infty} [(j^2 - k^2)u_{jk} - \beta k u_{j,-k}] \phi_{jk},$$

where $\Omega =]0, \pi[\times]0, 2\pi[$ and $\{\phi_{jk}\}$ denotes the orthonormal basis

$$\phi_{jk} = \begin{cases} \frac{\sqrt{2}}{\pi} \sin(jx) \sin(kt), & j \in \mathbb{Z}_+, k \in \mathbb{Z} \\ \frac{1}{\pi} \sin(jx), & j \in \mathbb{Z}_+, k = 0 \\ \frac{\sqrt{2}}{\pi} \sin(jx) \cos(kt), & j \in \mathbb{Z}, -k \in \mathbb{Z}_+ \end{cases}$$

and $u_{jk} = \langle u, \phi_{jk} \rangle$. The spectrum of the complex operator A_C is $\sigma(A_C) = \{\mu_{jk} = (j^2-k^2)+i\beta k \mid j \in \mathbb{Z}_+, k \in \mathbb{Z}\}$. It is easy to see that $\lambda_0 = \mu_{1,0} = 1$ and $\lambda_0 = \mu_{2,0} = 4$ are the only eigenvalues for which the condition ($\mathcal{A}$) holds with $\underline{\theta} = 0$ and $\overline{\theta} = 3$ or $\underline{\theta} = 3$ and $\overline{\theta} = 5$, respectively. Since Ker $A = \{\overline{0}\}$ we can apply Theorem 2(a) to produce

Theorem 3. Assume that $g = g(x,t,s)$ satisfies the conditions (g_1) and (g_3) with $\lambda_0 = 1$, $0 < a < 1 < b < 3$ or with $\lambda_0 = 4$, $3 < a < 4 < b < 5$. Then the telegraph equation (TE) admits at least two nontrivial weak solutions.

In order to obtain existence results also in other cases we need reductions to some subspaces H in $L^2(\Omega)$. This demands of course restriction for g to meet the invariance condition (g_2). To this end we consider the case where $g(x,t,s)$ is independent on x. We shall look for subspaces H such that the spectrum of the restriction of A to $H \cap D(A)$ satisfies $(\mathcal{A})$ for some eigenvalue λ_0. Indeed, for any real eigenvalue $\lambda_0 = j^2$ $(j \in \mathbb{Z}_+)$ we can find such a subspace. (cf. [4])

Theorem 4. Assume that $g = g(t,s)$ is odd in s and satisfies the conditions (g_1) and (g_3) with $\lambda_0 = j_0^2$ and $j_0^2 - 1 < a < j_0^2 < b < j_0^2 + 1$ for some $j_0 \in \mathbb{Z}_+$. Then the telegraph equation (TE) admits at least one pair $\{u, -u\}$ of nontrivial solutions.

Proof. We define

$$H = \{u \in L^2(\Omega) \mid u(x + \frac{2\pi}{j_0}, t) = u(x,t) \text{ a.e. } t \in]0, 2\pi[,$$
$$0 \leq x \leq \pi - \frac{2\pi}{j_0}, u(x,t) = -u(\frac{2\pi}{j_0} - x, t) \text{ a.e. } t \in]0, 2\pi[,$$
$$0 \leq x \leq \frac{2\pi}{j_0}\} = \overline{sp}\{\phi_{jk} \mid \frac{j}{j_0} \in \mathbb{Z}_+, k \in \mathbb{Z}\}$$

Then the complex spectrum of the operator $A|_H$ is $\{\mu_{qj_0,k} = (q^2 j_0^2 - k^2) + i\beta k \mid q \in \mathbb{Z}_+, k \in \mathbb{Z}\}$. Since $q = 1$, $k = 0$ is the only solution of the equation $q^2 j_0^2 - k^2 = j_0^2$, we can conclude that $A|_H$ satisfies the condition ($\mathcal{A}$) with

$$\underline{\theta} = \max\{q^2 j_0^2 - k^2 \mid q^2 j_0^2 - k^2 < j_0^2\} = j_0^2 - 1$$

and

$$\overline{\theta} = \min\{q^2 j_0^2 - k^2 \mid q^2 j_0^2 - k^2 > j_0^2\} \geq j_0^2 + 1.$$

Since $N(u(x,t)) = g(t, u(x,t))$, the invariance condition (g_2) also holds and the assertion follows from Theorem 2.

Remarks. 1) It is clear that the condition $b < j_0^2 + 1$ in Theorem 4 can be replaced by $b < \overline{\theta}$, where, of course, $\overline{\theta} \geq j_0^2 + 1$ depends on j_0. For instance, if $j_0 = 3$, $\overline{\theta} = 13$ and if $j_0 = 4$, $\overline{\theta} = 17$, respectively.
2) It is essential in theorem 4 that the nonlinearity g depends on t. Indeed, if $u \in D(A)$ is a weak solution of (TE) with $g = g(x, u)$, then it is easy to see that $u = u(x)$, and hence we would obtain solutions for the equation

$$\begin{cases} -u_{xx} = g(x, u) \\ u(0) = u(\pi) = 0. \end{cases}$$

References

[1] H. Amann, Saddle points and multiple solutions of differential equations, Math. Z. 169 (1979), 127-166.
[2] H. Amann and E. Zehnder, Multiple periodic solutions for a class of nonlinear autonomous wave equations, Houston J. Math. 7 (1981), 147-174.
[3] J. Berkovits and V. Mustonen, An extension of Leray-Schauder degree and applications to nonlinear wave equations, Differential and Integral Equations Vol. 3, No.5 (1990), 945-963.
[4] J. Berkovits and V. Mustonen, On the existence of multiple solutions for semilinear equations with monotone nonlinearities crossing a finite number of eigenvalues, Nonlinear Analysis T.M.A Vol. 17, No. 5 (1991), 399-412.

[5] H. Brézis and L. Nirenberg, Characterization of the ranges of some nonlinear operators and applications to boundary value problems, Ann. Scuola Norm. Sup. Pisa Ser. IV, 5 (1978), 225-326.

[6] N. P. Các, On a boundary value problem with nonsmooth jumping nonlinearity, J. Diff. Equations. 93 (1991), 238-259.

[7] S. Fučik and J. Mawhin, Generalized periodic solutions of nonlinear telegraph equations, Nonlinear Analysis, TMA Vol. 2, No. 5, (1978), 609-617.

[8] P. Hess, Nonlinear perturbations of linear elliptic and parabolic problems at resonance: Existence of multiple solutions, Ann. Scuola Norm. Sup. Pisa, Ser. IV, 5 (1978), 527-537.

[9] G. Hetzer, A spectral characterization of monotonicity properties of normal linear operators with an application to nonlinear telegraph equation, J. Operator Theory, 11 (1984), 333-341.

[10] A. C. Lazer and P. J. McKenna, Multiplicity results for a class of semilinear elliptic and parabolic boundary value problems, J. Math. Anal. Appl. 107 (1985), 371-395.

[11] A. C. Lazer and P. J. McKenna, Critical point theory and boundary value problems with nonlinearities crossing multiple eigenvalues, Comm. Partial Diff. Equations 11 (1986), 1653-1676.

[12] J. Mawhin, Periodic solutions of nonlinear telegraph equations, Dynamical Systems (ed. by Bednarek), Academic Press, New York (1977), 59-64.

[13] J. Mawhin, Nonlinear functional analysis and periodic solutions of semilinear wave equations, Nonlinear Phenomena in Mathematical Sciences (ed. by Lakshimikantham), Academic Press, New York (1982), 671-681.

Address

University of Oulu, Department of Mathematical Sciences, FIN-90570 Oulu, Finland.

L BOCCARDO
The role of truncates in nonlinear Dirichlet problems in L^1

1. Introduction and hypotheses

In some recent papers the existence of solutions for nonlinear elliptic equations with right hand side measures has been proved (see the references). The method used consists in approximating the right hand side with smooth functions, and then studying the behaviour of the solutions of the approximate equations (see Theorem 1.1, below). The aim of these notes is to show a strong convergence property of the truncations of these approximating solutions and then to prove the above existence result in the special case of right hand side positive summable functions. The main tools of this approach are the *a priori* estimate in the Sobolev space $W_0^{1,q}(\Omega)$ (see [BG1], [BG2], and [BGV]; the method of the latter paper will be used) and the almost everywhere convergence of the gradients of approximate equations (see [BG1], [BG2], [BM]) which will be proved using the strong convergence of the truncations of these approximating solutions, following [BO] (see also [LM], [BM]).

We begin stating the hypotheses that will hold throughout the paper.

Let Ω be a bounded, open subset of $\mathbf{R}^N$, $N \geq 2$. Let p be a real number such that $1 < p < N$.

Suppose that one of the following holds:

H1) Let $a : \Omega \times \mathbf{R}^N \to \mathbf{R}^N$ be a Carathéodory function such that, for almost every $x \in \Omega$, for every ξ and η in $\mathbf{R}^N$, with $\xi \neq \eta$,

$$\bigl(a(x,\xi) - a(x,\eta)\bigr) \cdot (\xi - \eta) > 0. \tag{1.1}$$

H2) Let p be such that $1 < p \leq 2$. Let $a : \Omega \times \mathbf{R} \times \mathbf{R}^N \to \mathbf{R}^N$ be a Carathéodory function such that for almost every $x \in \Omega$, for every s and t in $\mathbf{R}$, and for every ξ and η in $\mathbf{R}^N$,

$$\bigl(a(x,s,\xi) - a(x,s,\eta)\bigr) \cdot (\xi - \eta) \geq \alpha \frac{|\xi - \eta|^2}{(1 + |\xi| + |\eta|)^{2-p}}, \tag{1.2}$$

$$|a(x,s,\xi) - a(x,t,\xi)| \leq \theta(|s - t|) |\xi|^{p-1}, \tag{1.3}$$

where α is a positive real number, and θ is a positive, non decreasing function such that $\theta(0) = 0$ and

$$\int_{0+} \frac{ds}{\theta(s)} = +\infty. \tag{1.4}$$

Moreover, we will assume that, in both cases,

$$a(x,s,\xi) \cdot \xi \geq \beta |\xi|^p, \tag{1.5}$$

and

$$|a(x,s,\xi)| \leq k(x) + \beta_1 |s|^{p-1} + \beta_2 |\xi|^{p-1}, \tag{1.6}$$

for almost every $x \in \Omega$, for every $s \in \mathbf{R}$, for every $\xi \in \mathbf{R}^N$ (with obvious meaning if a does not depend on s), where β, β_1, β_2 are positive constants, and k belongs to $L^{p'}(\Omega)$.

Let us define the differential operator

$$\mathcal{A}(u,v) = -\mathrm{div}\,(a(x,u,\nabla v)), \tag{1.7}$$

thus making a little abuse of notation if the function a does not depend on s (i.e., if hypothesis H1 holds), and define $A(u) = \mathcal{A}(u,u)$. Thanks to (1.5) and (1.6), A is a pseudomonotone and coercive differential operator acting between $W_0^{1,p}(\Omega)$ and $W^{-1,p'}(\Omega)$; hence, it is surjective (see [11]).

We define, for s and k in $\mathbf{R}$, with $k \geq 0$, $T_k(s) = \max(-k,\min(k,s))$.

We recall the following result (see [BG1] and [BG3]).

Theorem 1.1. *Let p be such that $2 - \frac{1}{N} < p < N$. Let A be as before, and let μ be a bounded measure on Ω. Let $\{f_n\}$ be any sequence of smooth function that converges to μ in the weak* topology of measures, and let u_n be, for every n in $\mathbf{N}$, the solution of the boundary value problem*

$$\begin{cases} A(u_n) = f_n & \text{in } \Omega, \\ u_n = 0 & \text{on } \partial\Omega. \end{cases}$$

Then for every $q < \tilde{p} = \frac{N(p-1)}{N-1}$ there exists a positive constant c_q (independent of n) such that

$$\|u_n\|_{W_0^{1,q}(\Omega)} \leq c_q \qquad \forall n \in \mathbf{N}. \tag{1.8}$$

Moreover, u_n converges to a function u that is a solution in the sense of distributions of the Dirichlet problem

$$u \in W_0^{1,q}(\Omega) : A(u) = \mu \qquad \forall q < \tilde{p}.$$

Remark 1.2. In the previous theorem, the assumption $2 - \frac{1}{N} < p$ implies that $q \geq 1$ and $\nabla u \in L^1(\Omega)$. In [BBGGPV] a "good" functional setting is presented in order to prove the above existence theorem for every p such that $1 < p < N$.

Remark 1.3. The choice of $T_k(u_n)$ as test function in the problem solved by u_n, together with (1.5), shows that the sequence $\{T_k(u_n)\}$ is bounded in $W_0^{1,p}(\Omega)$, and so converges weakly to $T_k(u)$ in the same space, even if u_n converges to u only in $W_0^{1,q}(\Omega), q < \tilde{p} < p$.

In this paper, we shall prove that, under additional hypotheses on μ, it is possible to construct a particular sequence $\{f_n\}$ such that the convergence of $T_k(u_n)$ is strong in $W_0^{1,p}(\Omega)$.

If $\mu \in L^1(\Omega)$, $\mu \geq 0$, our result is the same of Theorem 5.1 of [BO], while if $\mu \in L^1(\Omega)$, without sign conditions, our result has been found by P.L. Lions and F. Murat in the context of renormalized solutions (see [LM]).

We note explicitely that our result does not imply that the truncates are strongly convergent for every sequence $\{f_n\}$ that weakly converges to μ. As has been shown in Remark 3.2 of [BM], there exists a sequence $\{f_n\}$, weakly convergent to μ, such that the corresponding sequence $\{T_k(u_n)\}$ is only weakly convergent in $W_0^{1,p}(\Omega)$.

2. The strong convergence result

We begin recalling the following comparison lemma, whose proof is straightforward.

Lemma 2.1. Let f_1, f_2 in $W^{-1,p'}(\Omega)$, with $0 \leq f_1 \leq f_2$, and let A be the differential operator defined by (1.7), where a satisfies hypothesis H1. Let u_1 and u_2 be the solutions of

$$\begin{cases} A(u_1) = f_1 & \text{in } \Omega, \\ u_1 = 0 & \text{on } \partial\Omega, \end{cases} \qquad \begin{cases} A(u_2) = f_2 & \text{in } \Omega, \\ u_2 = 0 & \text{on } \partial\Omega. \end{cases}$$

Then $0 \leq u_1 \leq u_2$ almost everywhere in Ω.

The following result is a modification of Proposition 1 of [AB] (see also [A]).

Lemma 2.2. Let f_1, f_2 in $W^{-1,p'}(\Omega)$, with $0 \leq f_1 \leq f_2$, and let A be the differential operator defined by (1.7), where a satisfies hypotheses H2. Let u_1 and

u_2 be the solutions of

$$\begin{cases} A(u_1) = f_1 & \text{in } \Omega, \\ u_1 = 0 & \text{on } \partial\Omega, \end{cases} \qquad \begin{cases} A(u_2) = f_2 & \text{in } \Omega, \\ u_2 = 0 & \text{on } \partial\Omega. \end{cases}$$

Then $0 \leq u_1 \leq u_2$ almost everywhere in Ω.

Proof. Subtract the equations satisfied by u_i, define $w = u_1 - u_2$, let ε be a positive real number, and choose as test function

$$v_\varepsilon = \int_0^{w^+} \frac{ds}{\theta_1(s+\varepsilon)^2}.$$

Since $A(u_1) - A(u_2) \leq 0$, and v_ε is positive, then

$$\int_\Omega \frac{(a(x,u_1,\nabla u_1) - a(x,u_2,\nabla u_2)) \cdot \nabla w^+}{\theta(s+\varepsilon)^2} \, dx \leq 0,$$

Setting $g(x) = 1 + |\nabla u_1| + |\nabla u_2|$ for the sake of simplicity, we obtain by (1.2), (1.3), and by Hölder inequality,

$$\alpha \int_\Omega \frac{|\nabla w^+|^2}{\theta(w^+ + \varepsilon)^2 \, g(x)^{2-p}} \, dx \leq \int_\Omega \frac{|\nabla w^+| |\nabla u_2|^{p-2}}{\theta(w^+ + \varepsilon)} \, dx$$

$$\leq \|u_2\|_{W_0^{1,p}(\Omega)}^{\frac{1}{p'}} \left(\int_\Omega \left| \frac{\nabla w^+}{\theta(w^+ + \varepsilon)} \right|^p dx \right)^{\frac{1}{p}}.$$

On the other hand, again by Hölder inequality,

$$\int_\Omega \left| \frac{\nabla w^+}{\theta(w^+ + \varepsilon)} \right|^p dx = \int_\Omega \left| \frac{\nabla w^+}{\theta(w^+ + \varepsilon) g(x)^{\frac{2-p}{2}}} \right|^p g(x)^{\frac{(2-p)p}{2}} \, dx$$

$$\leq \left(\int_\Omega \frac{|\nabla w^+|^2}{\theta_2(w^+ + \varepsilon)^2 g(x)^{2-p}} \, dx \right)^{\frac{p}{2}} \left(\int_\Omega g(x)^p \, dx \right)^{\frac{2-p}{2}}.$$

Hence, since both u_1 and u_2 are in $W_0^{1,p}(\Omega)$, we have

$$\int_\Omega \left| \frac{\nabla w^+}{\theta(w^+ + \varepsilon)} \right|^p dx \leq c_1 \left(\int_\Omega \left| \frac{\nabla w^+}{\theta(w^+ + \varepsilon)} \right|^p dx \right)^{\frac{1}{2}}.$$

By Poincaré inequality, this implies that

$$\int_\Omega \left(\int_0^{w^+} \frac{ds}{\theta(s+\varepsilon)} \right)^2 dx \leq c_2$$

45

which contradicts (1.4) as ε tends to zero if the set $\{x \in \Omega : w^+(x) \geq \rho\}$ has positive measure for some $\rho > 0$. Hence, $w^+ = 0$ almost everywhere in Ω, and so $u_1 \leq u_2$. □

The following lemma is an easy generalization of a result due to G. Stampacchia (see [S]).

Lemma 2.3. Let $u \in W_0^{1,p}(\Omega)$ be a positive supersolution of the boundary value problem
$$\begin{cases} A(u) = 0 & \text{in } \Omega, \\ u = 0 & \text{on } \partial\Omega. \end{cases} \quad (2.1)$$
Then $T_k(u)$ is a supersolution of the same problem for every $k \geq 0$.

Proof. Let k be a fixed positive number, and define the convex set
$$K = \{v \in W_0^{1,p}(\Omega) : v \geq T_k(u)\}.$$

Since $u \in K$, K is non-empty. Let z be a solution of the variational inequality
$$\begin{cases} z \in K, \\ \langle A(u,z), v - z \rangle \geq 0 \quad \forall v \in K, \end{cases} \quad (2.2)$$

where $\langle \cdot, \cdot \rangle$ is the duality pairing between $W^{-1,p'}(\Omega)$ and $W_0^{1,p}(\Omega)$. Choosing as test function $v = \min(z, u) \in K$, we obtain, by monotonicity and by the fact that z solves (2.2),
$$0 \leq \langle A(u,z) - A(u,v), z - v \rangle \leq \langle A(u,v), v - z \rangle = -\langle A(u,u), (u-z)^- \rangle \leq 0,$$

since u is a supersolution of (2.1). Hence, $z = v$, which implies that $T_k(u) \leq z \leq u$ almost everywhere in Ω. Thus, since $u = z$ where $u = T_k(u)$,
$$0 \leq \langle A(u,z) - A(u,T_k(u)), z - T_k(u) \rangle \leq \langle A(u,T_k(u)), T_k(u) - z \rangle$$
$$= \langle A(u,k), k - z \rangle = 0,$$

and so $z = T_k(u)$. Choosing $v = z + \varphi$ as test function in (2.2), with $\varphi \in W_0^{1,p}(\Omega)$, $\varphi \geq 0$, we obtain
$$\langle A(u,z), \varphi \rangle \geq 0 \quad \forall \varphi \in W_0^{1,p}(\Omega), \; \varphi \geq 0,$$

so that $A(u,z)$ is a positive distribution. Since $z = T_k(u)$, then $A(u, T_k(u)) = A(T_k(u))$ is a positive distribution, i.e. $T_k(u)$ is a supersolution of (2.1). □

We are ready to state and prove our strong convergence result.

Theorem 2.4. Let f be a _positive function_ in $L^1(\Omega)$, and let $\{f_n\}$ be an increasing sequence of
bounded functions such that

$$f_n \geq 0 \quad \forall n \in \mathbf{N}, \qquad f_n \to f \quad \text{strongly in } L^1(\Omega).$$

If u_n is, for every $n \in \mathbf{N}$, the solution of

$$\begin{cases} A(u_n) = f_n & \text{in } \Omega, \\ u_n = 0 & \text{on } \partial\Omega, \end{cases} \qquad (2.3)$$

then there exists a function $u \in W_0^{1,q}(\Omega)$, where q is any real number such that $1 \leq q < \tilde{p} = \frac{N(p-1)}{N-1}$, such that, for every $k \geq 0$,

$$T_k(u_n) \to T_k(u) \quad \text{strongly in } W_0^{1,p}(\Omega).$$

Proof. Choosing $f_n = T_n(f)$, we obtain an increasing sequence of positive bounded functions converging to f in $L^1(\Omega)$. From Lemma 2.1, or Lemma 2.2, it follows that

$$0 \leq u_n \leq u_{n+1} \qquad \forall n \in \mathbf{N}.$$

Now we prove that the sequence u_n is bounded in $W_0^{1,q}(\Omega)$ for any $q < \tilde{p}$. Let $\varphi_k(t)$ be the real valued function defined by

$$\varphi_k(t) = \begin{cases} 0 & \text{if } t \in [0, k], \\ t - k & \text{if } t \in (k, k+1], \\ 1 & \text{if } t \in (k+1, +\infty), \\ -\varphi_k(-t) & \text{otherwise.} \end{cases} \qquad (2.4)$$

Choosing $\varphi_k(u_n)$ as test function in the weak formulation of the approximate Dirichlet problem (2.3) yields

$$\alpha \int_{B_k} |\nabla u_n|^p \leq \int_\Omega a(x, u_n, \nabla u_n) \nabla \varphi_k(u_n) = \int_\Omega f_n \varphi_k(u_n) \leq \|f\|_{L^1(\Omega)} \qquad (2.5)$$

where

$$B_k = \{x \in \Omega : k \leq u_n(x) < k+1\}. \qquad (2.6)$$

Thus, for every $\lambda > 1$, we have

$$\int_\Omega \frac{|\nabla u_n|^p}{(1+u_n)^\lambda} = \sum_{k=0}^\infty \int_{B_k} \frac{|\nabla u_n|^p}{(1+u_n)^\lambda} \leq \sum_{k=0}^\infty \frac{\|f\|_{L^1(\Omega)}}{\alpha(1+k)^\lambda} = c_1 \qquad (2.7)$$

Let $q < p$; by Sobolev embedding and Hölder inequality,

$$c_0 \left(\int_\Omega (u_n)^{q*} \right)^{\frac{q}{q*}} \leq \int_\Omega |\nabla u_n|^q = \int_\Omega \frac{|\nabla u_n|^q}{(1+u_n)^{\frac{\lambda q}{p}}} (1+u_n)^{\frac{\lambda q}{p}} \leq$$

$$\left(\int_\Omega \frac{|\nabla u_n|^p}{(1+u_n)^\lambda} \right)^{\frac{q}{p}} \left(\int_\Omega (1+u_n)^{\frac{\lambda q}{p-q}} \right)^{1-\frac{q}{p}} \leq (c_1)^{\frac{q}{p}} c_2 \left(\int_\Omega (u_n)^{\frac{\lambda q}{p-q}} \right)^{1-\frac{q}{p}}$$

The previous inequalities give an a priori estimate of the sequence u_n in the large Sobolev space $W_0^{1,q}(\Omega), q < \tilde{p}$, if
(i) $\frac{q}{q*} > 1 - \frac{q}{p} \iff p < N$ (if $p > N$, $L^1(\Omega) \subset W^{1,p'}(\Omega)$)
(ii) $q^* = \frac{\lambda q}{p-q}$ and $\lambda > 1 \iff q < \tilde{p}$.

Hence, the above a priori estimate and Remark 1.2 imply that there exists a function $u \in W_0^{1,q}(\Omega)$ (for every q such that $1 \leq q < \tilde{p}$) such that

$$u_n \to u \qquad \text{weakly in } W_0^{1,q}(\Omega), \forall q < \tilde{p},$$

and

$$T_k(u_n) \to T_k(u) \qquad \text{weakly in } W_0^{1,p}(\Omega), \forall k \geq 0.$$

Since $\{u_n\}$ is an increasing sequence, so it is $\{T_k(u_n)\}$; hence, $T_k(u_n) \leq T_k(u)$ for every n in $\mathbf{N}$. Moreover, since $A(u_n) = \mathcal{A}(T_k(u_n), T_k(u_n)) \geq 0$, Lemma 2.3 implies that

$$A(T_k(u_n)) \geq 0. \qquad (2.8)$$

Hence,

$$0 \leq \langle \mathcal{A}(T_k(u_n), T_k(u_n)) - \mathcal{A}(T_k(u_n), T_k(u)), T_k(u_n) - T_k(u) \rangle$$
$$\leq -\langle \mathcal{A}(T_k(u_n), T_k(u)), T_k(u_n) - T_k(u) \rangle,$$

and the latter term tends to 0 as n tends to infinity since $\mathcal{A}(T_k(u_n), T_k(u))$ converges to $\mathcal{A}(T_k(u), T_k(u))$ strongly in $W^{-1,p'}(\Omega)$. Thus,

$$\lim_{n \to \infty} \langle \mathcal{A}(T_k(u_n), T_k(u_n)) - \mathcal{A}(T_k(u_n), T_k(u)), T_k(u_n) - T_k(u) \rangle = 0,$$

so that, by a result on strong convergence (see, e.g., [LL] or [BMP]), $T_k(u_n)$ converges to $T_k(u)$ strongly in $W_0^{1,p}(\Omega)$. □

Now we shall give a simple proof of the existence Theorem 1.1 if the right hand side is a *positive function* in $L^1(\Omega)$.

Theorem 2.5. *Let f be a positive function in $L^1(\Omega)$ and $2 - \frac{1}{N} < p < N$. Then there exists a function u, such that*

$$T_k(u) \in W_0^{1,p}(\Omega), \qquad (2.9)$$

for every $k > 0$, which is a weak solution of the Dirichlet problem

$$u \in W_0^{1,q}(\Omega) : A(u) = \mu \qquad \forall q < \tilde{p}. \qquad (2.10)$$

Moreover

$$\int_\Omega a(x, u, \nabla u)\, \nabla T_k(u - v) = \int_\Omega f\, T_k(u - v) \qquad (2.11)$$

for every $k > 0$ and every $v \in W_0^{1,p}(\Omega) \cap L^\infty(\Omega)$.

Proof. Let $\{u_n\}$ be the sequence defined in (2.3). In the previous theorem we have proved the existence of a subsequence, still denoted $\{u_n\}$, and of a function u in $W_0^{1,q}(\Omega)$, $q < \tilde{p}$, such that $T_k(u) \in W_0^{1,p}(\Omega)$, for any $k > 0$ and

$$u_n \rightharpoonup u \quad \text{weakly in } W_0^{1,q}(\Omega),\ q < \tilde{p}, \qquad (2.12)$$

$$T_k(u_n) \to T_k(u) \quad \text{strongly in } W_0^{1,p}(\Omega),\ \forall k > 0. \qquad (2.13)$$

Then, for every $r < q < \tilde{p}$, we have

$$\int_\Omega |\nabla(u_n - u)|^r \le c_1 \int_\Omega |\nabla T_k(u_n) - \nabla T_k(u)|^r + c_1 \int_{\{|u_n|>k\}} |\nabla u_n|^r + c_1 \int_{\{|u|>k\}} |\nabla u|^r$$

The first integral of the right hand side converges to zero by (2.13) and the fact that $r < p$. The last two terms are uniformly small with respect to n for k large enough. Then

$$u_n \to u \quad \text{strongly in } W_0^{1,r}(\Omega),\ \forall r < \tilde{p}, \qquad (2.14)$$

which implies that

$$a(x, u_n, \nabla u_n) \to a(x, u, \nabla u) \quad \text{strongly in } (L^s(\Omega))^N,\ \forall s < \frac{N}{N-1} \qquad (2.15)$$

Hence it is possible to pass to the limit in (2.3) in order to obtain that u is a weak solution of the Dirichlet problem (2.10).

Moreover, for every $v \in W_0^{1,p}(\Omega) \cap L^\infty(\Omega)$ and for every $h > 0$,

$$\nabla T_h(u_n - v) \to \nabla T_h(u - v) \quad \text{in measure},$$

by (2.13). Since

$$|\nabla T_h(u_n - v)| \leq |\nabla T_k(u_n)| + |\nabla v|, \quad k = h + \|v\|_{L^\infty(\Omega)},$$

the Vitali convergence theorem implies that

$$\nabla T_h(u_n - v) \to \nabla T_h(u - v) \quad \text{strongly in } W_0^{1,p}(\Omega), \forall h > 0. \tag{2.16}$$

The use of the convergences (2.15) and (2.16) in (2.3) yields, for $j = k + \|v\|_{L^\infty(\Omega)}$,

$$\int_\Omega a(x, u_n, \nabla u_n) \nabla T_k(u_n - v) = \int_\Omega a(x, T_j(u_n), \nabla T_j(u_n)) \nabla T_k(u_n - v) \to$$

$$\to \int_\Omega a(x, T_j(u), \nabla T_j(u)) \nabla T_k(u - v) = \int_\Omega a(x, u, \nabla u) \nabla T_k(u - v)$$

and

$$\int_\Omega f_n T_k(u_n - v) \to \int_\Omega f T_k(u - v)$$

which imply equality (2.10). □

Remark 2.6. We point out that, thanks to the strong convergence of truncates (2.13), we have proved equality (2.10), for a positive and summable function f. For every $f \in L^1(\Omega)$ and for monotone differential operators A, in [BBGGPV], it is proved the inequality

$$\int_\Omega a(x, \nabla u) \nabla T_k(u - v) \leq \int_\Omega f T_k(u - v) \tag{2.17}$$

for every $k > 0$ and every $v \in W_0^{1,p}(\Omega) \cap L^\infty(\Omega)$. However the inequality (2.17) is enough in order to prove the uniqueness of the weak solutions which satisfy it.

Remark 2.7. In the proof of Theorem 2.4, the existence of an increasing sequence $\{f_n\}$ plays a fundamental rôle (see also again Remark 1.2)

Remark 2.8. By (2.7) and (2.13) it follows that

$$A(T_k(u)) \geq 0, \qquad \forall k > 0. \tag{2.18}$$

On the other hand

$$\lim_{k \to \infty} \|u - T_k(u)\|_{W_0^{1,q}(\Omega)} = 0 \qquad \forall q < \tilde{p},$$

which implies that

$$A(T_k(u)) \to A(u) \qquad \text{strongly in } W^{-1,s}(\Omega), \ \forall s < \frac{N}{N-1}. \tag{2.19}$$

Then the positivity of the sequence $A(T_k(u))$ which is convergent in $W^{-1,s}(\Omega)$ implies that

$$A(T_k(u)) \to A(u) \qquad \text{in the weak} * \text{ topology of measures.} \tag{2.20}$$

See [BM2] for the proof of the above convergence in the general case.

Remark 2.9. For any $\lambda > 0$, consider the solutions $u_{\lambda,n}$ of the boundary value problem

$$\begin{cases} A(u_{\lambda,n}) + f^- \frac{u_{\lambda,n}}{\lambda + u_{\lambda,n}} = f_n^+, \\ u_{\lambda,n} \in W_0^{1,p}(\Omega), \ u_{\lambda,n} \geq 0. \end{cases}$$

It is possible to use the method of the previous Theorems 2.4 and 2.5 in order to prove that, for any $\lambda > 0$,

$$\lim_{n \to \infty} u_{\lambda,n} = u_\lambda \qquad \text{in } W_0^{1,q}(\Omega), \ q < \tilde{p},$$

where u_λ is the solution of

$$\begin{cases} A(u_\lambda) + f^- \frac{u_\lambda}{\lambda + u_\lambda} = f^+, \\ u_\lambda \in W_0^{1,q}(\Omega), \ q < \tilde{p}, \ u_\lambda \geq 0. \end{cases}$$

The behaviour (with respect to λ) of this kind of perturbated equations has been studied in [BC] and [BG2] in the framework of unilateral problems in $L^1(\Omega)$.

References

[A] M. Artola, *Sur une classe de problèmes paraboliques quasi-linéaires*, Boll. Un. Mat. Ital., **5** (1986), pp. 51–70.

[AB] M. Artola, L. Boccardo, *Nontrivial solutions for some quasilinear elliptic equations*, preprint.

[BBGGPV] Ph. Benilan, L. Boccardo, T. Gallouet, R. Gariepy. M. Pierre, J.L. Vazquez, *An L^1 theory of existence and uniqueness of solutions for nonlinear elliptic equations*, Ann. Scuola Norm. Sup. Pisa, to appear.

[BC] L. Boccardo, G.R. Cirmi, *Nonsmooth unilateral problems*, in Proceedings on nonsmooth optimization.

[BG1] L. Boccardo, T. Gallouet, *Nonlinear elliptic and parabolic equations involving measure data*, J. Funct. Anal., **87** (1989), pp.149–169.

[BG2] L. Boccardo, T. Gallouet, *Problèmes unilatéraux in L^1*, C. R. Acad. Sci. Paris, **311** (1990), pp. 617–619.

[BG3] L. Boccardo, T. Gallouet, *Nonlinear elliptic equations with right hand side measures*, Comm. Partial Differential Equations, **17** (1992) pp. 641–655.

[BG4] L. Boccardo, T. Gallouet, *Strongly nonlinear elliptic equations having natural growth terms and L^1 data*, Nonlinear Anal., **19** (1992), pp. 573–579.

[BG5] L. Boccardo, T. Gallouet, *Summability of the solutions of nonlinear elliptic equations with right hand side measures*, Convex Anal., to appear.

[BGMa] L. Boccardo, T. Gallouet, P. Marcellini, *Anisotropic equations in L^1*, Differential Integral Equations, to appear.

[BGMu] L. Boccardo, T. Gallouet, F. Murat, *A unified presentation of two existence results for problems with natural growth*, in Progress in PDE, the Metz surveys 2, M. Chipot editor, Research Notes in Mathematics, Longman, **296** (1993), pp. 127–137.

[BGO] L. Boccardo, T. Gallouet, L. Orsina, *Existence and uniqueness of entropy solutions for nonlinear elliptic equations with right hand side measures*, preprint.

[BGV1] L. Boccardo, T. Gallouet, J.L. Vazquez, *Some regularity results for some nonlinear parabolic equations in L^1*, Rend. Sem. Mat. Univ. Pol. Torino, fascicolo speciale (1989), pp. 69–74.

[BGV2] L. Boccardo, T. Gallouet, J.L. Vazquez, *Nonlinear elliptic equations in $\mathbf{R}^N$ without growth restrictions on the data*, J. Differential Equations, **105** (1993), 334–363.

[BM] L. Boccardo, F. Murat, *A property of nonlinear elliptic equations with the source term a measure*, Potential Analysis, **3** (1994), pp. 257–264.

[BMP] L. Boccardo, F. Murat, J.P. Puel, *Existence of bounded solutions for nonlinear elliptic unilateral problems*, Ann. Mat. Pura Appl., **152** (1988), pp. 183–196.

[BO] L. Boccardo, L. Orsina, *Semilinear elliptic equations in L^s*, Houston Journal of Mathematics, **20** (1994), pp. 99–114.

[D] A. Dall'Aglio, *Approximated solutions of equations with L^1 data. Application to the H-convergence of quasi-linear parabolic equations*, Ann. Mat. Pura e Appl., to appear.

[DO1] A. Dall'Aglio, L. Orsina, *Existence results for some nonlinear parabolic equations with nonregular data*, Differential Integral Equations, 5 (1992), pp. 1335–1354.

[DO2] A. Dall'Aglio, L. Orsina, *Nonlinear parabolic equations with natural growth conditions and L^1 data*, Nonlinear Anal., to appear

[DV] T. Del Vecchio, *Nonlinear elliptic equations with measure data*, Potential Analysis, to appear.

[LL] J. Leray, J.L. Lions, *Quelques résultats de Višik sur les problèmes elliptiques semi-linéaires par les méthodes de Minty et Browder*, Bull. Soc. Math. France, **93** (1965) pp. 97–107.

[LM] P.L. Lions, F. Murat, *Sur les solutions renormalisées d'équations elliptiques non linéaires*, to appear.

[MO] A. Malusa, L. Orsina, *Existence and regularity results for relaxed Dirichlet problems with measure data*, Ann. Mat. Pura Appl., to appear.

[O] L. Orsina, *Solvability of linear and semilinear eigenvalue problems with L^1 data*, Rend. Sem. Mat. Univ. Padova, **90** (1993), pp. 207–238.

[S] G. Stampacchia, *Équations elliptiques du second ordre à coefficients discontinus* Les Presses de l'Université de Montréal, Montréal (1966).

Dipartimento di Matematica, Università di Roma I, Piazza A. Moro 2, 00185, Roma, Italia
tel: (39.6)49913202; e-mail: boccardo@gpxrme.sci.uniroma1.it

A BOUCHERIF AND S M BOUGUIMA
Periodic solutions of second ordinary differential equations with a discontinuous nonlinearity

1 Introduction

There is a vaste literature on periodic boundary value problems for nonlinear second order ordinary differential equations. Most of the published work deals with sufficient conditions on the nonlinearity in order to have solvability. There are however several papers investigating necessay and sufficient conditions for nonresonance of periodic boundary value problems (see for instance [3], [4], [5], and the references therein). In this paper we are interested in the solvability of the following periodic boundary value problem

$$-x'' = g(x) + h(t) \qquad t \in I: = [0, 2\pi]$$

$$x(0) - x(2\pi) = x'(0) - x'(2\pi) = 0$$

where $h \in L^\infty(I; R)$ and $g: R \to R$ satisfies some conditions that will be specified later. Our objective is to extend the main result in [5] to the case of a discontinuous nonlinearity.

2 The assumptions

Let S denote the set of $(\lambda, \mu) \in R^2$ such that the piecewise linear problem

$$-x'' = \lambda x^+ - \mu x^- \quad \text{in} \quad I, \ x(0) - x(2\pi) = x'(0) - x'(2\pi) = 0 \qquad (2)$$

has a nontrivial solution. It was shown by Fucik that $S = \bigcup_{n=0}^{\infty} C_n$ with

$$C_0 = \{(\lambda, \mu) \in R^2; \ \lambda, \mu = 0\},$$

and

$$C_n = \{(\lambda, \mu) \in R^2; \ \lambda \rangle n^2/4, \ \mu \rangle n^2/4 \ \text{and} \ \lambda^{-1/2} + \mu^{-1/2} = 2n^{-1}\}$$

(see [4, chap.42] and [3, p.22])

We shall assume that the nonlinearity g is continuous everywhere except at 0 and satisfies

(H1) $g(0-) = \lim_{\varepsilon \to 0^+} g(0-\varepsilon) < g(0+) = \lim_{\varepsilon \to 0^+} g(0+\varepsilon)$ and $g(0) \in [g(0-), g(0+)]$

(H2) $\limsup_{s \to +\infty} \frac{g(s)}{s} \leq \lambda$ and $\limsup_{s \to -\infty} \frac{g(s)}{s} \leq \mu$

(H3) $\limsup_{s \to +\infty} \frac{2G(s)}{s^2} < \lambda$ and $\limsup_{s \to -\infty} \frac{2G(s)}{s^2} < \mu$

where $(\lambda, \mu) \in C_1$ and $G(s) = \int_0^s g(u)\,du$.

Remarks. 1. The case $g(0+) < g(0-)$ can be treated similarly
2. If $(\lambda, \mu) \in C_n$ for $n > 1$, a multiplicity result has been obtained in [2] for a problem similar to (1).

3 The main result

Theorem. Assume that, in addition to (H1), (H2), (H3), there exist two real numbers A and B such that

$$-h(t) \in [g(A), g(B)] \cap (R \smallsetminus [g(0-), g(0+)]) \quad \text{a.e. in I} \qquad (3)$$

Then (1) has a solution x in the following sense:

$$-x'' - h(t) \in \tilde{g}(x) \quad \text{a.e. in I}, \quad x(0) - x(2\pi) = x'(0) - x'(2\pi) = 0 \qquad (4)$$

has a solution, where

$$\tilde{g}(x) = \{g(x)\} \text{ if } x \neq 0 \text{ and } \tilde{g}(0) = [g(0-), g(0+)]$$

Remarks. 3. This result is due to Gossez and Omari for continuous g (see [5])
4. We can assume g discontinuous at some other point.

Proof. Assume first $A \geq B$. Then (4) has a solution. In fact, if $A = B$ then $x(t) = A$ is a solution of (4).

Now, if $A > B$ we can suppose $A > 0 > B$. We will reduce the solvability of (4) to a fixed point problem.

Let $K: L^2(I;R) \to H^2(I;R)$ be the compact operator such that Kp is the unique solution of
$$-x'' + x = p,\ x(0) - x(2\pi) = x'(0) - x'(2\pi) = 0$$

Let Γ $(L^2(I;R))$ denote the set of all nonempty, closed convex subsets of $L^2(I;R)$. We define a multifunction
$$F: H^1(I;R) \to \Gamma(L^2(I;R)) \quad \text{by}$$

$$F(x) = \begin{cases} \{g(x) + h(t)\} & \text{if } x \neq 0 \\ [g(0-) + h(t),\ g(0+) + h(t)] & \text{if } x = 0 \end{cases}$$

Then problem (4) has a solution x if and only if

$$x \in KF(x) \tag{5}$$

First, note that F is upper semicontinuous (see [1]).
Next, for $\lambda \in [0,1]$ consider the following equation

$$x \in \lambda K(x + g(x) + h(t)) \text{ a.e. in } I \tag{6}$$

Claim 1. There exists a constant c_1, independent of λ and x, such that

$$\|x'\|_{L^2} \leq c_1 \text{ for all } x, \text{ solution of (4) satisfying}$$

$$A \geq x(t) \geq B \quad \forall t \in I$$

Now, if $-x'' - h(t) \in \tilde{g}(x)$ and $I_o = \{t \in I;\ x(t) = 0\}$

Then $I_o = \phi$, for otherwise we would have $-h(t) \in [g(0-), g(0+)]$ which is impossible. Hence $-x''-h(t) = g(x)$ a.e. in I.
Then, from (6), we have

$$-x'' = -(1-\lambda)x + \lambda(g(x) + h(t)) \quad \text{a.e. in } I \tag{7}$$

Equation (7) yields

$$\int_0^{2\pi}(x')^2 dt = \lambda \int_0^{2\pi} g(x)xdt + \lambda \int_0^{2\pi} h(t)xdt - (1-\lambda)\int_0^{2\pi} x^2 dt$$

$$= \lambda \int_{I_1} g(x)xdt + \lambda \int_{I_2} g(x)xdt + \lambda \int_0^{2\pi} h(t)xdt - (1-\lambda)\int_0^{2\pi} x^2 dt \le c_1$$

where

$$I_1 = \{t \in I;\ 0 \le x(t) \le A\} \quad \text{and} \quad I_2 = \{t \in I;\ B \le x(t) \le 0\}$$

Claim 2. Let

$$O_1 = \{x \in H^1(I; R);\ \|x'\|_{L^2} \le c_1 \text{ and } A > x(t) > B\ \forall t \in I\}$$

Then (6) has no solution $x \in \partial O_1$ for any $\lambda \in [0,1]$.
For suppose on the contrary that there exists $x \in \partial O_1$ solution of (6). Now, $x \in \partial O_1$ means min $x = B$ or max $x = A$. Let $t_1 \in I$ be such that min $x = x(t_1) = B$. Then $x'(t_1) = 0$. On the other hand there exists $\varepsilon > 0$ small enough such that

$$-(1-\lambda)B + \lambda(g(B) + h(t)) \ge \varepsilon \quad \text{a.e. in } I$$

It follows, from the continuity of x, that

$$-(1-\lambda)x(t) + \lambda(g(x) + h(t)) > 0 \quad \text{a.e. in } I$$

in some neighborhood of t_1. This leads to a contradiction, and the claim is proved.

If max x = A we proceed in a similar way. Therefore, by the Leray degree for multifunctions (see [1]) we can conclude that (5) is solvable, i.e. KF has a fixed point, which is a solution of (4) and consequently a solution of (1).

Next, if $A < 0 < B$ we can assume $g(s)\text{sgns} \geq -c_2$ for all $s \in R$, where c_2 is some positive constant. We will adapt the proof of theorem 2.1 in [5] to the case of a discontinuous nonlinearity.

Let θ be a real number such that $0 < \theta < q_+$ and $0 < \theta < q_-$, where $(q_+, q_-) \in C_1$. Then the periodic problem

$$-x'' - \theta x = e, \quad x(0) - x(2\pi) = x'(0) - x'(2\pi) = 0$$

has a unique solution x, which we denote by $K_2\, e(t)$. This defines an operator K_2.
Problem (1) is then reduced to the fixed point problem

$$x \in K_2(-\theta x + g(x) + h) \quad \text{in } H^1(I; R) \tag{8}$$

The homotopic equation

$$x \in \lambda K_2(-\theta x + g(x) + h) \quad \text{for } \lambda \in [0,1] \tag{9}$$

corresponds to the equation

$$-x'' \in (1-\lambda)\theta x + \lambda[g(x) + h] \quad \text{a.e. in } I \tag{10}$$

Let $I_1 = \{t \in I;\ x(t) = 0\}$. Then, it follows from (3) that meas $(I_1) = 0$.
We have the following claims.

Claim 3. There exists a positive constant c_3 (independent of x and λ) such that if x is a solution of (10), then

$$\max x \leq c_3 \quad \text{or} \quad \min x \geq -c_3$$

Proof. see claim 1 in [5].

Claim 4. There exists a positive constant c_4 (independent of x and λ) such that if x is a solution of (10), then

$$\|x'\|_{L^2} \leq c_4$$

To see this, let x be a solution of (10) for some λ.
Set $\phi(x,\lambda) := (1-\lambda)\theta x + \lambda g(x)$. Then, there exists a constant c such that

$$\left|\int_0^{2\pi} \phi(x(t),\lambda)dt\right| \leq c$$

We have that either $\max x \leq c_3$ or $\min x \geq -c_3$. Consider the first case, the second is similar. We have

$$\int_0^{2\pi} |\phi(x(t),\lambda)|dt = \int_{I_1} |\phi(x(t),\lambda)|dt + \int_{I \setminus I_1} |\phi(x(t),\lambda)|dt$$

Note that
$$\int_{I_1} |\phi(x(t),\lambda)|dt = \lambda g(0) \operatorname{meas}(I_1) = 0$$

Now on $I \setminus I_1$ g is continuous and so the proof in [5] goes through. Hence we have

$$\int_0^{2\pi} |\phi(x(t),\lambda)|dt \leq c$$

Consequently, the estimate on $\|x'\|_{L^2}$ follows.

Finally, let

$$O_2 = \left\{ x \in H^1(I;R); \; \|x'\|_{L^2} < c+1 \;\; \text{and} \;\; A < x(t) < B \;\; \text{for at least one } t \right\}.$$

Then O_2 is a bounded open set in $H^1(I;R)$ which contains 0. Now proceed as in the case $A \geq B$ to complete the proof.

Research partially supported by a MESRS-DRS Grant B1329/03/01/89

References

[1] K.C.Chang, Free boundary problems and set - valued mapping, J.Diff. Eqns. 49 (1983), 1-28

[2] M.A.Del Pino, R.F. Manasevich and A. Murua, On the number of 2π- periodic solutions for u" + g (u) = s(1+h (t) using the Poincaré-Birkhoff Theorem, J.Diff.Eqns. 95 (1992), 240-258.

[3] P. Drabek, Solvability and Bifurcation of Nonlinear Equations, Pitman Research Notes in Math. 264, Longman, 1992.

[4] S. Fucik, Solvability of Nonlinear Equations and Boundary Value Problems, D.Reidel Publ. Co, Holland 1980.

[5] J.P. Gossez and P.Omari, Periodic solutions of a second order O.D.E.: a necessary and sufficient condition for nonresonace J. Diff. Eqns. 94 (1991), 67-82.

A. Boucherif and S. M. Bouguima
Department of Mathematics
University of Tlemcen
B.P. 119 Tlemcen, 13000 Algeria

The range of sums of monotone operators and applications to Hammerstein inclusions and nonlinear complementarity problems

1 Introduction.

Several differential inclusions and optimization problems are given involving perturbations of nonlinear operators. The equations studing these problems are of the type

$$find\ u \in X\ such\ that\ f \in Au + Bu. \tag{1}$$

When A and B are two monotone operators an interesting method which deal with this inclusion problem is to consider the range of their sum. In general the sum of their ranges, $R(A) + R(B)$, is much large than the range of their sum $R(A+B)$. Nevertheless, it has been seen in many cases that these sets are almost equal in the sense that they have the same closures and interiors, i.e. $R(A) + R(B) \approx R(A+B)$. There is an abstract literature on this problem. Since the pioneering works [5] of Brézis and [7] of Brézis and Haraux, who obtain existence in (1) without using coercivness assumption, there have been several attempts to generalize these results from a Hilbert space. In [7] Brézis and Haraux deal with the range of the sum of monotone operators in a Hilbert space. Some of their results have been extended to accretive operators in L^p, for $1 < p < +\infty$, and reflexive Banach spaces by Browder [9], Gupta [16], Gupta and Hess [17], Calvert and Gupta [11], and by Reich [27] in some restrictive Banach spaces. For monotone operators, some results in reflexive Banach spaces have been established by Riech [27]. For other extensions and generalizations to non monotone operators see the papers of Brézis and Nierenberg [8] and Gupta [16]. Also, several recent results involving ranges of sums of accretive operators and compact perturbations have been treated in Banach spaces by Kartsatos [19] [20] [21], Morales [22]. In all these articles, geometrical properties of uniform smoothness of Banach spaces are assumed.

Such results on the range of sums can be applied to several problems. Brézis [5] and Brézis & Haraux [7] give some applications for Hammerstein equations and to some nonlinear boundary value problems.

The purpose of this paper is to further generalize Brézis and Haraux theorem for monotone operators to nonreflexive Banach spaces, as similarly established in [26], and to give applications of this results to some nonlinear problems. The paper is organized as follows. In the subsequent section 2 we introduce some preliminaries on the maximal monotone operators and the densely monotone operators introduced by Gossez, and collect some technical tools. Whereas section 3 gives the main reults of this paper. In the final section 4 the paper concludes with two applications. The first deals with minimization of the sum of convex functions, the second one is devoted to Hammerstein differential inclusions in Banach spaces, and the last one treats the complementarity problems for monotone operators in reflexive Banach spaces.

2 Preliminaries.

Let us begin by recalling some definitions which can be found in [7], [13], [24], [25] and [29].

We suppose throughout that X is a real Banach space. Let X^* be its topological dual space and denote by $\langle .,. \rangle$ the canonical bilinear form between X and X^*. We use $cl_\tau(C)$ (or $\overline{C}^\tau$) and $int_\tau(C)$ to denote, respectively, the closure and the interior of a subset C of X relatively to the topology τ. When τ is the strong (norm) topology, we write $\overline{C}$ and $int(C)$. We shall denote in X and X^* by $a + rB$ the closed ball centred at $a \in X$ with radius $r > 0$. The convex hull of C will denoted by $co\, C$.

An operator $A : D \subset X \longrightarrow 2^{X^*}$ is a multivalued mapping from X to X^*. Occasionally we shall identify an operator A with its graph : $\xi \in A(x)$ is equivalent to $(x, \xi) \in A$. The domain, the range and the inverse of A are defined by $D(A) := \{x \in D\ :\ A(x) \neq \emptyset\}$, $R(A) := \{x^* \in X^*\ :\ x^* \in A(x)\ for\ some\ x \in D(A)\}$ and $A^{-1}(x^*) := \{x \in X\ :\ x^* \in A(x)\}$.

The pointwise sum of two operators A and B is $(A+B)(x) := A(x) + B(x) = \{x^* + y^*\ :\ x^* \in A(x)\ and\ y^* \in B(x)\}$.

2.1 Maximal monotone operators.

Let us recall that an operator $T : D \subset X \longrightarrow 2^{X^*}$ is said to be monotone if $\forall x, y \in D(A), \forall \xi \in A(x)$ and $\forall \eta \in A(y)$, $\langle \xi - \eta, x - y \rangle \geq 0$.

A monotone operator A is called maximal monotone in D provided that $(x, \xi) \in D \times X^*$ belongs to A if and only if $\langle \xi - \eta, x - y \rangle \geq 0\ \forall y \in D(A), \forall \eta \in A(y)$.

Let $\epsilon \geq 0$, the ϵ-subdifferential of an extended convex real function $f : X \longrightarrow \mathbf{R} \cup \{+\infty\}$ at a point x of its domain $dom(f)$, i.e. $f(x) < +\infty$, is defined by

$$\partial_\epsilon f(x) := \{x^* \in X^*\ :\ f(x) + f^*(x^*) - \langle x^*, x \rangle \leq \epsilon\}$$

where $f^*(x^*) := \sup_{y \in X}(\langle x^*, y \rangle - f(y))$ is the Fenchel conjugate function.

If f is a proper convex lower semicontinuous function, the operator $\partial f := \partial_0 f$, see [13] or [24] for arecent and interesting proof established by Simons, is a maximal monotone operator.

The ϵ-approximate duality mapping of X, as introduced by Gossez [13], is given by $J_\epsilon(x) := \partial_\epsilon(\frac{1}{2}\|.\|^2)(x)$. If $\epsilon = 0$ we have $J(x) := J_0(x) = \{x^* \in X^*\ :\ \langle x^*, x \rangle = \|x\|.\|x^*\| = \|x\|^2\}$ is the duality mapping of X.

Lemma 2.1 ([13]) *Let $\epsilon \geq 0$. Then*
(a) $x^* \in J_\epsilon(x)$ *implies that* $|\|x\| - \|x^*\|| \leq \sqrt{2\epsilon}$ *and* $\langle x^*, x \rangle \geq \|x\|.\|x^*\| - \epsilon$;

(b) *the operator J_ϵ is coercive, i.e.* $\displaystyle\lim_{\|x\|\to+\infty} \frac{\langle J_\epsilon(x), x\rangle}{\|x\|} = +\infty.$

Definition 2.1 *An operator A is said to be 3^*-monotone if A is monotone and*

$$\inf_{(y,y^*)\in A} \langle x^* - y^*, x - y\rangle > -\infty \quad \forall x \in D(A), \forall x^* \in R(A). \tag{2}$$

An interesting example of operators satisfying the 3^*-monotonicity in a general Banach space is the subdifferential of proper convex lower semicontinuous functions. Other strong properties like *p-cyclic monotonicity* for $p \geq 3$ and *angle-boundedness* have been introduced. For survey articles about this class of operators we cite the papers [5]-[7], [29].

2.2 Densely monotone operators.

As inserted by Gossez [13], let us introduce the definition of some monotone operators for which the notion of maximality have been extended.

Let τ_o be the weakest topology on X^{**} which renders continuous the following real functions

$$\begin{cases} X^{**} \longrightarrow \mathbf{R} \\ x^{**} \longmapsto \langle x^{**}, x^*\rangle \end{cases} \forall x^* \in X^* \text{ and } \begin{cases} X^{**} \longrightarrow \mathbf{R} \\ x^{**} \longmapsto \|x^{**}\| \end{cases}$$

The topology τ we shall consider in $X^{**} \times X^*$ is the product topology of τ_o and the strong (norm) topology τ_s of X^*.

If X is normed, but not complete, X^{**} is certainly large than X, since it is complete. But even if X is a Banach space X^{**} is generally large than X. Here we identify the space X with a subspace of X^{**}, and hence the norm induced on X by X^{**} is the norm of X. By the same we identify the bilinear form between X and X^*, and between X^* and X^{**} for elements $x^* \in X^*$ and $x \in X$, i.e. $_{X^*}\langle x^*, x\rangle_X =_{X^{**}}\langle x, x^*\rangle_{X^*}$.

Definition 2.2 *An operator $A : X \longrightarrow 2^{X^*}$ is said to be densely monotone iff the operator $cl_\tau A : X^{**} \longrightarrow 2^{X^*}$ is maximal monotone.*
Here $cl_\tau A = \overline{A}^\tau$ is the operator which graph is the closure of the graph of A, relatively to the topology τ.

Remarks. (i) If X is supposed to be reflexive, then every maximal monotone operator is densely monotone, and $\overline{A}^\tau = A$.
The converse does not hold in general. It suffies to take $X = \mathbf{R}$, $D(A) = \mathbf{R}\setminus\{0\}$ and $A(x) = 0 \; \forall x \in D(A)$.

(ii) If f is a proper convex lower semicontinuous, then its subdifferential ∂f is densely monotone and $\overline{\partial f}^\tau = \partial^{-1} f^* := (\partial f^*)^{-1}$.
We finally give, as obtained by Gossez [13], some basic properties of densely monotone operators.

Theorem 2.1 ([13], Lemme 2.1) *Let $A : X \longrightarrow 2^{X^*}$ be a densely monotone operator. Then $(x^{**}, x^*) \in \overline{A}^\tau$ if and only if $\langle x^{**} - y, x^* - y^* \rangle \geq 0$ for every $(y, y^*) \in A$.*

Theorem 2.2 ([13], Théorème 4.1) *Let $A : X \longrightarrow 2^{X^*}$ be a densely monotone operator. Then $R(\lambda J_\epsilon + A) = X^* \quad \forall \lambda > 0, \ \forall \epsilon > 0$.*

3 Main results.

Now we drive the main result of this paper :

Theorem 3.1 *Let $A, B : X \longrightarrow 2^{X^*}$ be two operators such that $A + B$ is densely monotone. Assume one of the following*
  (i) *A and B are 3^*-monotone, or*
  (ii) *A is 3^*-monotone and $D(B) \subset D(A)$.*
Then

$$\overline{R(A+B)} = \overline{R(A) + R(B)} \text{ and } int(R(A+B)) \subset int(R(A)+R(B)) \subset int(R(\overline{A+B}^\tau)).$$

Proof. It suffies to show that $\overline{R(A+B)} \supset \overline{R(A) + R(B)}$ and $int(R(A) + R(B)) \subset int(R(\overline{A+B}^\tau))$; other inclusions are trivial.
Let us take $C = A + B$ and $S = R(A) + R(B)$.

Step 1. For each $x^* \in S$ there exists $x \in X$ such that

$$\inf_{(y,y^*) \in C} \langle x^* - y^*, x - y \rangle > -\infty. \tag{3}$$

Indeed, let $z^* \in S$. Then there exist $x^* \in R(A)$ and $y^* \in R(B)$ such that $z^* = x^* + y^*$. If (i) is satisfied, let $x \in D(A) \cap D(B)$. From 3^*-monotonocity of A and B we obtain

$$\inf_{(u,u^*) \in C} \langle z^* - u^*, x - u \rangle \geq \inf_{(u,v^*) \in A, (u,w^*) \in B} (\langle x^* - v^*, x - u \rangle + \langle y^* - w^*, x - u \rangle)$$
$$\geq \inf_{(u,v^*) \in A} \langle x^* - v^*, x - u \rangle + \inf_{(u,w^*) \in B} \langle y^* - w^*, x - u \rangle > -\infty.$$

If (ii) is satisfied, let $x \in D(B)$ such that $y^* \in B(x)$. $D(B) \subset D(A)$ implies $x \in D(A)$. From monotonicity of B and 3^*-monotonocity of A we obtain

$$\inf_{(u,u^*) \in C} \langle z^* - u^*, x - u \rangle \geq \inf_{(u,v^*) \in A} \langle x^* - v^*, x - u \rangle + \inf_{(u,w^*) \in B} \langle y^* - w^*, x - u \rangle$$
$$\geq \inf_{(u,v^*) \in A} \langle x^* - v^*, x - u \rangle > -\infty.$$

Step 2. Let us prove that $S \subset \overline{R(C)}$.
Let $x^* \in S$. Since $C = A + B$ is densely monotone, from Theorem 2.2, for a fixed $\epsilon > 0$ and every $\lambda > 0$ there exist $x_\lambda \in D(C)$, $y^*_\lambda \in J_\epsilon x_\lambda$ and $z^*_\lambda \in C(x_\lambda)$ such that $x^* = \lambda y^*_\lambda + z^*_\lambda$.

FFrom (3), there exist a real $c > 0$ and $x \in X$ such that $\langle x^* - u^*, x - u \rangle \geq -c \ \forall (u, u^*) \in C$. Let us take $u = x_\lambda$ and $u^* = z_\lambda^* = x^* - \lambda y_\lambda^*$, then $\langle z_\lambda^* - x^*, x_\lambda - x \rangle = \langle \lambda y_\lambda^*, x - x_\lambda \rangle \geq -c$.

Since $y_\lambda^* \in J_\epsilon(x_\lambda)$, we have $\langle y_\lambda^*, x_\lambda \rangle \geq \|y_\lambda^*\|^2 + \|x_\lambda\|^2 - \epsilon$. Hence by elementary algebraic calculations and using relations above, we get that $\lambda \|x_\lambda\|^2 \leq \lambda \|x\|^2 + 2\lambda\epsilon + 2c$. We deduce that $(\sqrt{\lambda}\|x_\lambda\|)_{\lambda > 0}$ is bounded relatively to λ. Using $\lambda \|y_\lambda^*\| \leq \lambda\sqrt{2\epsilon} + \lambda \|x_\lambda\|$ and $z_\lambda^* = x^* - \lambda y_\lambda^* \in C(x_\lambda)$, it follows that $x^* \in \overline{R(C)}$.

Step 3. Let us prove that $int(S) \subset R(\overline{C}^\tau)$.
Let x^* belong to $int(S)$, there exists then $r > 0$ such that $x^* + rB \subset S$.
By (3), with $z_\lambda = x^* - \lambda y_\lambda^* \in C(x_\lambda)$ we have for some $c > 0$,

$$\forall \lambda > 0, \ \langle x^* - \lambda y_\lambda^* - x^* - z^*, x_\lambda - z \rangle = \langle -\lambda y_\lambda^* - z^*, x_\lambda - z \rangle \geq -c.$$

Fix $\lambda_0 > 0$ and take $\lambda \in]0, \lambda_0]$, since $y_\lambda^* \in J_\epsilon(x_\lambda)$, we have

$$\langle z^*, x_\lambda \rangle \leq c + \langle z^*, z \rangle + \langle \lambda y_\lambda^*, z - x_\lambda \rangle \leq c + \langle z^*, z \rangle + \frac{\lambda_0}{2}\|z\|^2 + \lambda_0 \epsilon - \frac{\lambda}{2}\|x_\lambda\|^2.$$

Then

$$\sup_{0 < \lambda \leq \lambda_0} \langle z^*, x_\lambda \rangle < +\infty.$$

The boundedness of the family $\{x_\lambda : 0 < \lambda \leq \lambda_0\}$ in X is achieved via use of Uniform Boundedness Principle. This implies that (x_λ) is also bounded in X^{**}. By Alaoglu's theorem there exists a subfamily also denoted by (x_λ) which converge, relatively to $\sigma(X^{**}, X^*)$ topology, to a point $\overline{x} \in X^{**}$. Let us prove that $\overline{x} \in D(\overline{C}^\tau)$ and that $x^* \in \overline{C}^\tau(\overline{x})$, which from Theorem 2.1, is equivalent to prove that

$$\langle x^* - y^*, \overline{x} - y \rangle \geq 0 \ \forall (y, y^*) \in C.$$

Let $(y, y^*) \in C$. Since $z_\lambda^* \in C(x_\lambda)$, by monotonicity of C, we have $\langle z_\lambda^* - y^*, x_\lambda - y \rangle \geq 0 \ \forall \lambda > 0$.
Therefore, since $z_\lambda^* - x^* = -\lambda y_\lambda^* \to 0$ and $x_\lambda \to \overline{x}$ for $\sigma(X^{**}, X^*)$, we obtain

$$0 \leq \limsup_{\lambda \searrow 0} \langle x_\lambda - y, z_\lambda^* - y^* \rangle$$
$$\leq \limsup_{\lambda \searrow 0} \langle x_\lambda, z_\lambda^* - x^* \rangle + \limsup_{\lambda \searrow 0} \langle x_\lambda, x^* - y^* \rangle + \limsup_{\lambda \searrow 0} \langle y, y^* - z_\lambda^* \rangle$$
$$\leq \langle \overline{x}, x^* - y^* \rangle + \langle y, y^* - x^* \rangle = \langle \overline{x} - y, x^* - y^* \rangle.$$

Thus $\langle \overline{x} - y, x^* - y^* \rangle \geq 0$, and this completes the proof. □

Corollary 3.1 *Suppose that the Banach space X is reflexive. Let A and B be two monotone operators in X such that either A and B are 3^*-monotone or A is 3^*-monotone and $D(B) \subset D(A)$. If $A + B$ is maximal, then $R(A + B) \approx R(A) + R(B)$, i.e. the interiors and the closures are the same.*

Proof. Apply Theorem 3.1 and use $A + B = \overline{A+B}^\tau$, since $A + B$ is maximal. □

Corollary 3.2 *Let X be a Banach space and f and g be two proper convex lower semicontinuous real functions. Suppose that $dom(f) \cap dom(g) \neq \emptyset$ and*

$$\overline{\bigcup_{t>0} t(dom(f) - dom(g))} \text{ is a closed linear subspace of } X \tag{4}$$

Then $\overline{R(\partial(f+g))} = \overline{R(\partial f) + R(\partial g)}$ and $int(R(\partial f) + R(\partial g)) = int(R(\partial^{-1}(f+g)^*))$.

Proof. Apply [1] Corollary 1, assumption (4) implies, if $A = \partial f$ and $B = \partial g$, that $A + B = \partial(f+g)$. Thus $A + B$ is densely monotone. From particular properties of the subdifferential mappings A and B are 3^*-monotone and $\overline{\partial(f+g)}^\tau = \partial^{-1}(f+g)^*$. Using Theorem 3.1 we conclude. □

Proposition 3.1 *If A is a densely monotone operator in a Banach space X, then $\overline{R(A)}$ is convex. Assume further that $\overline{A}^\tau = A$, then $int\ R(A)$, if it is nonempty, is convex also.*

Proof. Let x^* belong to $R(A)$, we have the existence of some $x \in D(A)$ such $x^* \in A(x)$. Using monotonicity of A, we have

$$\inf_{(y,y^*) \in A} \langle x - y, x^* - y^* \rangle > -\infty. \tag{5}$$

Thus each point in co $R(A)$, also satisfies (5). The result follows by applying this observation and the proof of Theorem 3.1 to $C = A$ and $S = co\ R(A)$. □

Remark. In general, the convexity of $\overline{R(A)}$ is not true for more general maximal monotone operator in general Banach space. Gossez, see [15], has already constructed a maximal monotone operator from l^1 to l^∞ whose range has not a convex closure.

Proposition 3.2 *Let A be a densely monotone operator in a Banach space X. Suppose that co $\overline{D(A)}$ is weakly compact, then $R(\overline{A}^\tau)$ is equal to the space X^*.*

Proof. Under similar proof as in [23] (p. 42), we can verify that property (3) is satisfied when $S = X^*$. So that conditions of Theorem 3.1 are abviously satisfied. Consequently $int\ S = X^* \subset R(\overline{A}^\tau)$. □

4   Applications.

In this section we present applications illustrating the interest of our work. For other applications one can consult the papers [7], [8] and [29].

4.1 Application to optimization problems.

Our first application deals with minimization of the sum of two convex functions. Let X be a Banach space and $f, g : X \longrightarrow R \cup \{+\infty\}$ be two proper convex lower semicontinuous functions. Let us suppose assumptions of Corollary 3.2 and that $0 \in int(R(\partial f) + R(\partial g))$. From Corollary 3.2 one has $0 \in R(\partial^{-1}(f+g)^*) = D(\partial (f+g)^*)$. Let $\bar{x} \in \partial (f+g)^*(0)$, then $(f+g)^*(0) + (f+g)^{**}(\bar{x}) = 0$. This implies that $(f+g)^{**}(\bar{x}) = \inf_{x \in X}(f(x) + g(x))$.

Note that $f+g$ is proper convex lower semicontinuous, so that if $\bar{x} \in X$, $(f+g)^{**}(\bar{x}) = (f+g)(\bar{x})$. We conclude that

$$(f+g)(\bar{x}) = \min_{x \in X}(f(x) + g(x)).$$

More generally, we have obtained the existence solution for every linear perturbation near the function $f+g$: $\exists r > 0$ such that if $x^* \in rB$, there exists $x \in dom(f) \cap dom(g)$ which satisfies

$$f(x) + g(x) + \langle x^*, x \rangle = min_X(f + g + \langle x^*, . \rangle)$$

4.2 Application to Hammerstein differential inclusions.

Here the second application deals with the Hammerstein differential inclusion in a Banach space. Let X be a Banach space, $F : D(F) = X^* \longrightarrow 2^X$ and $G : D(G) = X \longrightarrow 2^{X^*}$ be two monotone operators. Let f be a fixed element of X^*. We consider the problem of finding $u \in X^*$ such that

$$f \in u + GF(u). \tag{6}$$

Clearly, this problem can be expressed in two forms as follows:

$$find \ u \in X^* \ such \ that \ \ 0 \in F(u) - G^{-1}(f - u). \tag{7}$$

or

$$find \ v \in X \ such \ that \ \ f \in F^{-1}(v) + G(v). \tag{8}$$

Theorem 4.1 *Let X be a Banach space. Suppose one of the operators F and G is 3^*-monotone and either*

(i) X is reflexive and F, G are maximal; or

(ii) X is a non-reflexive Banach space and $F^{-1} + G$ is maximal in $X^{**} \times X^*$.

Then for each $f \in X$ the problem (6) has a solution.

Proof. Consider the operators $A = F^{-1}$ and $B = G$. So that A and B are monotone. Assumption (i) and (ii) implies that those of Theorem (3.1) are satisfied. Thus

$$R(\overline{A + B^\tau}) = R(\overline{F^{-1} + G^\tau}) = R(F^{-1} + G) \supset int(R(F^{-1}) + R(G)) = int(D(F) + R(G))$$
$$= X^*.$$

We conclude from (8) that the Hammerstein inclusion has a solution. □

One can formulate an example of this problem by considering the nonlinear integral equation of Hammerstein :

$$u(x) + \int_\Omega g(x,y)f(y,u(y))dy = h(x) \quad a.e. \ on \ \Omega \qquad (9)$$

where Ω is a measurable subspace of R^N, dx is the Lebesgue measure on Ω, g is the real kernel function defined on $\Omega \times \Omega$ and f is a real function on $\Omega \times \mathbf{R}$.
Let us consider the following assumptions on f :

(1) f is a Carathéodory's function, i.e. for all $t \in \mathbf{R}$, $f(.,t)$ is measurable, and for a.e. $x \in \Omega$, $f(x,.)$ is continuous.
(2) f is monotone, i.e. $t_1 \leq t_2 \Leftrightarrow f(x,t_1) \leq f(x,t_2)$ for a.e. $x \in \Omega$.
(3) For all $t \in \mathbf{R}$, $f(.,t)$ lies in $L^1(\Omega; \mathbf{R})$.

Let us define the two operators on $v \in L^p(\Omega; \mathbf{R})$ and $w \in L^q(\Omega; \mathbf{R})$ by

$$G(v)x := \int_\Omega g(x,y)v(y)dy \quad and \quad F(w)x := f(x,w(x)).$$

The second operator F is usually called the Niemytski operator associated to f.
On G we suppose the following assumption :

(4) For all $v \in L^1(\Omega; \mathbf{R})$ $\langle G(v), v \rangle := \int\int_\Omega g(x,y)v(x)v(y)dxdy \geq 0.$

We take $X = L^p(\Omega; \mathbf{R})$ for $1 < p \leq +\infty$, and $X^* = L^q(\Omega; \mathbf{R})$ where $q = \dfrac{p}{p-1}$.
Suppose that $h \in X^*$. Under these data, the Hammerstein inetgral equation can be written as $u + GFu = h$.

In order to justify existence of solutions of the Hammerstein integral equation, let us verify conditions of Theorem 4.1.
Assumptions (1) and (2) imply that Niemytski operator F is well defined on $X^* = L^q(\Omega; \mathbf{R})$ and is monotone.
Assumptions (2) and (3) justify that F maps elements from $L^\infty(\Omega; \mathbf{R})$ to $L^1(\Omega; \mathbf{R})$.
Moreover, from Carathéodory conditions, the operator F is continuous on $L^\infty(\Omega; \mathbf{R})$.
One can verify also, see [6], that F is 3^*-monotone since it is 3-cyclically monotone.
Let us prove the two assumptions (i) and (ii) of Theorem 4.1 from which the result of existence follows.

Existence when $1 < p < +\infty$. In this case, we have F and G are maximal monotone. Since $D(G) = X$ we have $X = D(G) - D(F^{-1})$ and we deduce maximality of $F^{-1} + G$. So that from Theorem 4.1, for every $h \in L^q(\Omega; \mathbf{R})$ there exists $v \in L^p(\Omega; \mathbf{R})$ such that $F^{-1}(v) + G(v) = h$.

Existence when $p = +\infty$. In this nonreflexive case, for conditions concerning the maximal monotonicity of $F^{-1} + G$ we refer to recent paper [12] of the authors where the principle theorem can be formulated as follows

Lemma 4.1 ([12], Theorem 2.3) *Let X be a Banach space, $T_1, T_2 : X \longrightarrow 2^{X^*}$ be two maximal monotone operators such that*
 (i) $\forall a \in D(T_1), \exists r > 0, \exists \rho > 0$ such that $\forall x \in a + \rho B \ T_1(x) \cap rB \neq \emptyset$;
 (ii) there exists a convex subset C of X such that $D(T_1) \subset C$ and $\forall x \in D(T_1)$ the contingent cone to C at x coinsides with the contingent cone to $D(T_1)$;
 (iii) $D(T_2) = X$.
Then $T_1 + T_2 : D(T_1 + T_2) = D(T_1) \to 2^{X^*}$ is maximal monotone.

Let us verify conditions of lemma. We have $D(G) = X$ and $D(F^{-1}) = R(F)$ is convex. For convexity of $R(F)$, let $v_i \in F(u_i)$ for $i = 1, 2$ and let $\lambda \in]0.1[$. By mean value theorem, we have for a.e. $x \in \Omega$ the existence of $u_\lambda(x) \in [u_1(x), u_2(x)]$ such that $\lambda F(u_1)x + (1 - \lambda)F(u_2)x = f(x, u_\lambda(x))$. Thus $u_\lambda \in L^1(\Omega; \mathbf{R})$ and $\lambda F(u_1) + (1 - \lambda)F(u_2) = F(u_\lambda)$, i.e. $u_\lambda \in R(F)$. This implies that conditions (ii) and (iii) of Lemma are fulfieled.
For condition (i) it suffies to use the assumption (2) on monotonicity of f. FFrom Lemma we conclude that $F^{-1} + G$ is maximal monotone in $L^1(\Omega; \mathbf{R}) \times L^\infty(\Omega; \mathbf{R})$.
Let us verify that $\overline{F^{-1} + G}^{\tau} = F^{-1} + G$.
Take $(u, v) \in \overline{F^{-1} + G}^{\tau}$, there is a net $(u_\alpha, v_\alpha)_{\alpha \in I}$ such that for each $\alpha \in I \ (u_\alpha, v_\alpha) \in F^{-1} + G$ such that (u_α, v_α) converges for topology τ to (u, v), that is $\int_\Omega u_\alpha(x)w(x)dx \to u(w)$ for all $w \in L^\infty(\Omega; \mathbf{R})$, $\|u_\alpha\|_{L^1} \to \|u\|_{(L^\infty)^*}$ and $v_\alpha \to v$ for the norm topology in $L^\infty(\Omega; \mathbf{R})$.
FFrom this we deduce that for each $\epsilon > 0$, there is $\delta > 0$ such that if the measure μ on Ω satisfies $\mu(E) < \delta$ then for every $\alpha \in I$, $\int_E |u_\alpha(x)|dx \leq \epsilon$, and hence $u(1_E) \leq \epsilon$.
If we set $\Psi(E) = u(1_E)$ and denote $(\Omega, \mathcal{A})$ the measurable space, we obtain a measure on $\mathcal{A}$ which is absolutly continuous with respect to the Lebesgue measure. So, we can find a function $\overline{u} \in L^1(\Omega; \mathbf{R})$ such that for all measurable ϕ one has

$$\int_\Omega \phi d\Psi = \int_\Omega \phi(x)\overline{v}(x)dx \text{ and } u(1_E) = \int_E \overline{v}(x)dx \text{ for all } E \in \mathcal{A}.$$

Then for every $w \in L^\infty(\Omega; \mathbf{R}) \ u(w) = \int_\Omega \overline{u}(x)w(x)dx$ and therefor $u_\alpha \to \overline{u}$ for $\sigma((L^\infty)^*, L^\infty)$. This implies that $u \in L^1(\Omega; \mathbf{R}))$ and $u_\alpha \to u$ for $\sigma(L^1, L^\infty)$.
Now from maximality of $F^{-1} + G$, the convergence of u_α to u for topology $\sigma(L^1, L^\infty)$ and v_α to v for norm topology in $L^\infty(\Omega; \mathbf{R})$, we deduce that $v \in (F^{-1} + G)(u)$.
This ends the proof of this assertion.

Using Theorem 4.1 we conclude the following result

Theorem 4.2 *Under assumptions (1)-(4) we conclude that the integral Hammerstein equation admits a solution.*

4.3 Application to complementarity problems.

As a final application we consider, in a reflexive Banach space X, the problem of finding $(x, x^*) \in X \times X^*$ such that

$$x \in P, x^* \in -P^* \text{ and } x^* \in A(x) \tag{10}$$

where $A : X \to 2^{X^*}$, P is a closed convex cone and $P^* := \{u^* \in X^*; \langle u^*, u \rangle \leq 0 \ \forall u \in P\}$ is the corresponding dual cone.

Theorem 4.3 *Suppose that A is 3^*-monotone and maximal monotone in X. If $\bigcup_{\lambda > 0} \lambda(D(A) - P)$ is a closed linear subspace of X and $0 \in int(R(A) - P^*)$, then complementarity problem (10) admits a solution.*

Proof. The complementarity problem (10) is equivalent to find $x \in P$ such that $0 \in \tilde{A}(x) := A(x) + \partial \delta_P(x)$, where $\delta_P(u) = 0$ if $u \in P$ and $\delta_P(u) = +\infty$ if $u \notin P$. For this it suffies to remark that $-x^* \in P^*, x \in P$ is similar to $x^* \in \partial \delta_P(x)$.
FFrom $0 \in int(R(A) - P^*)$, we have $0 \in int(R(\tilde{A}) + R(\partial \delta_P))$. Since $\bigcup_{\lambda > 0} \lambda(D(A) - P)$ is a closed subspace of X, we deduce from [2] that $A + \partial \delta_P$ is maximal monotone. Applying Corollary 3.1 we obtain a solution from $0 \in int(R(\tilde{A} + \partial \delta_P)) = int(R(A + \partial \delta_P)) = int(R(\tilde{A}))$. $\square$

Remark. Condition $0 \in int(R(A) - P^*)$ is satisfied when $int(P^*) \cap R(A) \neq \emptyset$ or $P^* \cap int(R(A)) \neq \emptyset$. In finite dimension space R^m, we always set $P = R_+^m$ and $A = \partial f$ where f is a convex lower semicontinuous function.
In this case $int P = \mathbf{R}_{++}^m = \{x \in \mathbf{R}^m : x_k > 0 \text{ for } k = 1, ..., m\}$ and $int P^* = -\mathbf{R}_{++}^m$.

References

[1] ATTOUCH, H. and H. BRÉZIS, *Duality for the sum of convex functions in general Banach spaces*, In "ASPECTS OF MATHEMATICS AND ITS APPLICATIONS", J.B. Arroso ed., North Holland, 1986, pp. 125-133.

[2] ATTOUCH, H., H. RIAHI and M. THÉRA *Somme ponctuelle d'opérateurs maximaux monotones*, Serdica Math. Journal, (to appear).

[3] AUBIN, J.-P., *L'Analyse Non Linéaire et ses Motivations Economiques*, Masson, 1984.

[4] AUBIN, J.-P. and I. EKELAND, *Applied Nonlinear Analysis*, Wiley, New York, 1984.

[5] BRÉZIS, H., *Quelques propriétés des opérateurs monotones et des semi-groupes non linéaires*, Wiley, New York, 1984, pp. 56-82.

[6] BRÉZIS, H. and F.E. BROWDER, *Nonlinear equations and systems of Hammerstein type*, Advances in Math., 18(1975), pp. 115-147.

[7] BRÉZIS, H. and A. HARAUX, *Image d'une somme d'opérateurs monotones et applications*, Isr. J. Math., 23(1976), pp. 165-186.

[8] BRÉZIS H. and L. NIERENBERG, *Characterization of the ranges of some nonlinear operators and applications to boundary value problems*, Ann. Scuola Norm. Sup. Pisa Cl.Sci., Ser. IV, 5(1978), pp. 225-326.

[9] BROWDER, F.E., *On the principle of H.Brézis and its applications*, J. Funct. Anal., 25(1977), pp. 356-365.

[10] BROWDER, F.E., *Image d'un opérateurs maximal monotone et le principe de Landesman-Lazer*, C.R. Acad. Sci.Paris, Série A, 287(1978), pp.715-718.

[11] CALVERT, B. and C. GUPTA, *Nonlinear elliptic boundary value problems in L^p space and sums of ranges of accretive operators*, Nonlinear Anal., T.M.A., 2(1978), pp. 1-26.

[12] CHBANI, Z. and H. RIAHI, *Sur la maximalité de la somme de deux opérateurs maximaux monotones définis sur un Banach général*, C.R. Acad. Sci.Paris, (submited).

[13] GOSSEZ, J.-P., *Opérateurs monotones non linéaires dans les espaces de Banach non réflexifs*, J. Math. Anal. Appl., 34(1971), pp. 371-395.

[14] GOSSEZ, J.-P., *On the range of a coercive maximal monotone operator in a nonreflexive Banach space*, Proc. Amer. Math. Soc., 35(1972), pp. 88-92.

[15] GOSSEZ, J.-P., *On the convex property of the range of a maximal monotone operator*, Proc. Amer. Math. Soc., 55(1976), pp. 359-9360.

[16] GUPTA, C.P., *Sum of ranges of operators and applications*, in "NONLINEAR SYSTEMS AND APPLICATIONS", Academic Press, New York, 1977 pp. 547-559.

[17] GUPTA, C.P. and P. HESS, *Existence theorems for nonlinear noncoercive operator equations and nonlinear elliptic boundary value problems*, J. Diff. Eq., 22(1976), pp. 305-313.

[18] KAPLAN, D.R., and A.G. KARTSATOS, *Ranges of sums and control of nonlinear evolutions with pressigned responses*, J. Optimization Theory and Appl., 81(1994), pp. 121-141.

[19] KARTSATOS, A.G., *Mapping theorems involving ranges of sums on nonlinear operators*, Nonlinear Anal., T.M.A., 6(1982), pp. 271-278.

[20] KARTSATOS, A.G., *On compact perturbations and compact resolvents of nonlinear m-accretive operators in Banach spaces*, Proc. Amer. Math. Soc., 119(1993), pp. 1189-1199.

[21] KARTSATOS, A.G., *Sets in the ranges of sums for perturbations on nonlinear m-accretive operators in Banach spaces*, Proc. Amer. Math. Soc., 123(1995), pp. 145-156.

[22] MORALES, C.H., *On the range of sums of accretive and continuous operators in Banach spaces*, Nonlinear Anal., T.M.A., 19(1992), pp. 1-9.

[23] PAVEL, N.H., *Analysis of some Nonlinear Problems in Banach Spaces and Applications*, Univ. Al.I.Cuza, Fac. Mat., 1982.

[24] PHELPS, R.R., *Convex Functions, Monotone operators and Differentiability*, Springer-Verlag, Berlin Hidelberg, 2nd edition 1993.

[25] PHELPS, R.R., *Lectures on maximal monotone operators*, Lectures given at Prague-Paseky Summer School, Czech-Republic, 1993, pp. 15-28.

[26] RIAHI, H., *On the range of the sum of two monotone operators in general Banach spaces*, Proc. Amer. Math. Soc. (to appear).

[27] REICH, S., *The range of sums of accretive and monotone operators*, J. Math. Analysis Appli., 68(1979), pp. 310-317.

[28] ROCKAFELLAR, R.T., *On the maximal monotonicity of subdifferential mappings*, Pacific J. Math., 33(1970), pp. 209-216.

[29] ZEIDLER, E., *Nonlinear Functinal Analysis and Applications II-B : Nonlinear Monotone Operators*, Springer-Verlag, New York, 1990.

ADDRESS

Université Cadi Ayyad - Faculté des Sciences I - Mathematics, BP. S 15, Boulevard My Adbellah, 40000 Marrakech, Morocco.

A priori estimates for positive solutions of semilinear elliptic systems via Hardy–Sobolev inequalities

0. Introduction

In this paper we establish a priori bounds for positive solutions of certain superlinear elliptic systems of the type

$$(S) \begin{cases} -\Delta u = f(x, u, v, Du, Dv) & \text{in } \Omega \subset \mathbf{R}^N, \\ -\Delta v = g(x, u, v, Du, Dv) & \text{in } \Omega \subset \mathbf{R}^N, \\ u = v = 0 & \text{on } \partial\Omega \end{cases}$$

and for a related polyharmonic equation, $m > 1$, i.e.

$$(P) \begin{cases} (-\Delta)^m u = h(x, u, \Delta u, \ldots, \Delta^{m-1} u) & \text{in } \Omega \subset \mathbf{R}^N, \\ u = \Delta u = \ldots = \Delta^{m-1} u = 0 & \text{on } \partial\Omega. \end{cases}$$

Here Ω is a smooth bounded domain of $\mathbf{R}^N$ with $N \geq 3$ and f, g and h are given functions that we will specify later.

When dealing with superlinear problems, the question of the existence of a priori bounds of positive solutions in L^∞ norm is of fundamental importance for proving existence results especially when the problem under consideration is not of variational type. In the case of scalar equations many techniques have been developed in the recent years for solving this problem. See for example [BT], [FLN],[GS] and [PL].

In particular for the prototype equation

$$\begin{cases} -\Delta u = f(x, u) & \text{in } \Omega \subset \mathbf{R}^N, \\ u = 0 & \text{on } \partial\Omega. \end{cases} \tag{0.1}$$

with $f(x, u) \simeq u^p$, the question of the a priori bounds admits a rather satisfactory answer. See [FLN] and [GS].

As it is well known, one of the main features of elliptic systems is the possible lack of variational structure. This property is not only important for the application of variational methods but also because it seems connected with the question of the a priori bounds in itself (see [FLN]). In the treatment of superlinear problems (i.e. $f(x, u) \simeq u^p$ with $p > 1$) an important concept related to existence of non trivial solutions is the notion of *criticality*.

This vague term has by now a sufficiently precise meaning in the context of scalar equations. It is in fact well accepted that one can call (0.1) *subcritical* if f satisfies a condition of the type

$$|f(x,u)| \leq C(|u|^p + 1), \text{ for every } (x,u) \in \Omega \times R \tag{0.2}$$

and $1 < p < \frac{N+2}{N-2}$ for $N \geq 3$. The number $\frac{N+2}{N-2}$ is the critical Sobolev exponent for the imbedding $H_0^1(\Omega) \hookrightarrow L^{p+1}(\Omega)$ and the fact that we require that $1 < p < \frac{N+2}{N-2}$ has many interpretations in different but related contexts concerning equation (0.1) for example : Palais-Smale conditions for the associated Euler-Lagrange functional, Liouville type theorems , classification of singular solutions, bootstrap procedures, etcetera.

In the case of general superlinear elliptic systems the concept of *criticality* is not yet well understood. Again this is due, roughly speaking , to the possible lack of variational structure or to the difficulty of understanding the coupling that takes place between the equations. To be more specific let us consider a typical model problem. Let $F : R^2 \to R$ be of class C^2 . We can consider two natural systems associated with F namely:

$$-\Delta u = \frac{\partial F}{\partial u}(u,v) \quad , \quad -\Delta v = \frac{\partial F}{\partial v}(u,v), \tag{0.3}$$

or

$$-\Delta u = \frac{\partial F}{\partial v}(u,v) \quad , \quad -\Delta v = \frac{\partial F}{\partial u}(u,v). \tag{0.4}$$

We shall call (0.3) a system of potential type, while (0.4) of Hamiltonian type. It can be shown that the concept of *criticality* for these two systems is essentially different. In particular for the canonical model

$$F(u,v) = \frac{1}{p+1}|v|^p v + \frac{1}{q+1}|u|^q u \tag{0.5}$$

where $1 < p, q$, the potential system is *subcritical* if

$$\max\{p,q\} < \frac{N+2}{N-2} \tag{0.6}$$

(observe that this is an uncoupled system), while the hamiltonian system (see [M$_1$], [vdV], [CFM], [PvdV]) is *subcritical* if

$$\frac{1}{p+1} + \frac{1}{q+1} > \frac{N-2}{N}. \tag{0.7}$$

For non-variational systems containing more than two equations the situation can become much more complicated (see for example [M$_2$]). This leaves open the question of the general understanding of *criticality* for superlinear elliptic systems.

We point out that in view of the results of Pucci and Serrin (see [PS]) on variational identities one can relate *criticality* of general variational systems with non-existence results of Pohozaev's type. However it is not yet clear if *subcriticality* together with some additional conditions is sufficient for a priori estimates of positive solutions.

We now come back to the question of the a priori bounds for our specific problem. A method for proving existence of a priori estimates of positive solutions of scalar second order elliptic equations has been proposed in 1977 by Brézis and Turner [BT]. Their method combines Hardy and Sobolev inequalities via interpolation and it is very powerful even when the problem under study is not variational.

The goal of this paper is to use Brézis-Turner method for finding a priori bounds for positive solutions of system (S) and for the higher order equation (P). We point out that an earlier result of this type using the Brézis-Turner method has been obtained by Cosner [C]. However, Cosner's result does not cover ours as simple example shows. (see Remark 2.1-(iv) in section 2 below).

This paper is organized as follows. Section 1 contains the preliminary information about Hardy-Sobolev's inequalities and some variations of them suitable for our purposes. In Section 2 we state and prove the main result of this paper concerning system (S). Finally in Section 3 we treat in detail the question of the a priori bounds of positive solutions for the Navier problem (P).

1. Hardy-Sobolev inequalities

Let Ω be a bounded smooth domain of $\mathbf{R}^N$ with $N \geq 3$ and let φ be the principal eigenfunction of $(-\Delta; H_0^1(\Omega))$, normalized by $\int_\Omega \varphi \, dx = 1$, with corresponding eigenvalue $\lambda_1 > 0$.

It is well known that for any $q > 1$ and $u \in W_0^{1,q}(\Omega)$ we have the following Hardy inequality

$$\left\|\frac{u}{\varphi}\right\|_{L^q} \leq C \|Du\|_{L^q} \tag{1.1}$$

where the constant C depends only on N and q. The case $q = 2$ has been proved in Lions-Magenes [LM ; p.59] while for general values of u the result is contained in Kavian [K]. Inequality (1.1) is a consequence of the following classical Hardy's result proved in the late 1920's.

" If $q > 1$, $f(x) \geq 0$ for any $x \in R_+$ and $F(x) = \int_0^x f(t) \, dt$ then

$$\int_0^{+\infty} \left(\frac{F(x)}{x}\right)^q dx < \frac{q}{q-1} \int_0^{+\infty} f(x)^q \, dx, \tag{1.2}$$

unless $f \equiv 0$."

The constant $\frac{q}{q-1}$ is the best possible, as it is shown for example in [HLP]. The extension of (1.2) to higher dimensions follows standard techniques of breaking the domain Ω into patches and mapping the boundary into half-spaces and then applying (1.2). Interesting extensions of (1.2) in the one dimensional case in more general form are contained in Ziemer [Z] and Mazja [MA].

Using (1.1) we can establish the following interpolation inequality. In what follows we make the convention that $\frac{1}{0} = \infty$.

Lemma 1.1. *Let $r_0 \in]1, \infty]$, $r_1 \in [1, \infty[$ and $u \in L^{r_0}(\Omega) \cap W_0^{1,r_1}(\Omega)$. Then for all $\tau \in [0,1]$ we have*

$$\frac{u}{\varphi^\tau} \in L^r(\Omega)$$

where

$$\frac{1}{r} = (1-\tau)\frac{1}{r_0} + \frac{\tau}{r_1}. \tag{1.3}$$

Moreover

$$\left\|\frac{u}{\varphi^\tau}\right\|_{L^r} \le C \|u\|_{L^{r_0}}^{1-\tau} \|u\|_{W^{1,r_1}}^{\tau}, \tag{1.4}$$

where the constant C depends only on τ, r_0 and r_1.

Proof. The inequality is trivially true if $\tau = 0$ while if $\tau = 1$ it reduces to (1.1). Let us suppose that $\tau \in (0,1)$. We have

$$\int_\Omega |u|^r \varphi^{-\tau r} \, dx = \int_\Omega |u|^{\tau r} \varphi^{-\tau r} |u|^{(1-\tau)r} \, dx, \tag{1.5}$$

hence if $r_0 < \infty$, by Hölder's inequality with $p = \dfrac{r_1}{r\tau}$ and $p' = \dfrac{r_0}{(1-\tau)r}$ we can estimate the right hand side of (1.5) with

$$\left(\int_\Omega \left|\frac{u}{\varphi}\right|^{r_1} dx\right)^{\frac{\tau r}{r_1}} \left(\int_\Omega |u|^{r_0} dx\right)^{\frac{(1-\tau)r}{r_0}}. \tag{1.6}$$

An application of (1.1) gives (1.4).

In case $r_0 = \infty$ by estimating (1.5) with

$$\|u\|_{L^\infty}^{(1-\tau)r} \left(\int_\Omega \left|\frac{u}{\varphi}\right|^{\tau r} dx\right) \tag{1.7}$$

and using again (1.1), with $q = r_1 = \tau r$, we achieve the proof.

There are some interesting special cases of (1.4) that we shall use later for our purposes.

Special cases of (1.4)

(i) - Let $u \in W_0^{1,q}(\Omega)$ with $N > q$. Choosing $r_1 = q$ and $r_0 = \dfrac{Nq}{N-q}$, by (1.4) and Sobolev imbedding we obtain

$$\left\|\frac{u}{\varphi^\tau}\right\|_{L^r} \leq C\|Du\|_{L^q}, \tag{1.8}$$

where

$$\frac{1}{r} = \frac{1}{q} - \frac{(1-\tau)}{N}. \tag{1.9}$$

The case $q = 2$ appears in Brézis-Turner [BT] and this more general case in Kavian [K].

(ii) - Let $u \in W_0^{1,q}(\Omega)$ with $N < q$. In this case $W_0^{1,q}(\Omega) \subset L^\infty(\Omega)$ and by Lemma 1.1 with $r_0 = \infty$ and $r_1 = q$ we have

$$\left\|\frac{u}{\varphi^\tau}\right\|_{L^r} \leq C\|u\|_{L^\infty}^{1-\tau} \|u\|_{W^{1,q}}^\tau. \tag{1.10}$$

Therefore by Sobolev's imbedding

$$\left\|\frac{u}{\varphi^\tau}\right\|_{L^r} \leq C\|Du\|_{L^q}, \tag{1.11}$$

where $\dfrac{1}{r} = \dfrac{\tau}{q}$.

(iii) - Let $u \in W_0^{1,q}(\Omega)$ with $N = q$. In this case we have $W_0^{1,q}(\Omega) \subset L^p(\Omega)$ for any $1 < p < \infty$. From Lemma 1.1 and Sobolev imbedding we see that (1.8) still holds for any $\dfrac{1}{r} > \dfrac{\tau}{q}$. We remark that in this case we do not recover (1.1) with $\tau = 1$. However, Kavian (see [K]) proved that also in this limit situation (1.8) holds with $\dfrac{1}{r} = \dfrac{\tau}{q}$.

The following result will be useful later.

Corollary 1.1. *Let $u \in W_0^{1,s}(\Omega) \cap W^{2,s}(\Omega)$ where $1 < s < \infty$. Assume that either*

$$\frac{1}{r} = \frac{1}{s} - \frac{(2-\tau)}{N} \quad \text{if} \quad s < \frac{N}{2} \tag{1.12}$$

or

$$\frac{1}{r} = \frac{\tau}{N} \quad \text{if} \quad s = \frac{N}{2} \tag{1.13}$$

or

$$\frac{1}{r} = \tau(\frac{1}{s} - \frac{1}{N}) \quad \text{if} \quad \frac{N}{2} < s < N \tag{1.14}$$

or
$$\forall r > 1, \forall \tau \in [0,1] \text{ if } N \leq s. \tag{1.15}$$

Then
$$\left\|\frac{u}{\varphi^\tau}\right\|_{L^r} \leq C \|u\|_{W^{2,s}}, \tag{1.16}$$

where the constant C depends only on τ, s and N.

Proof. We shall proceed by distinguishing the various cases under consideration.

(a) - Let $s < \frac{N}{2}$ and let r_0 and r_1 be such that

$$\frac{1}{r_0} = \frac{1}{r_1} - \frac{1}{N} = \frac{1}{s} - \frac{2}{N}. \tag{1.17}$$

It follows that
$$W^{2,s}(\Omega) \subset W^{1,r_1}(\Omega)$$
and
$$W^{2,s}(\Omega) \subset L^{r_0}(\Omega).$$

Applying Lemma 1.1 with r given by (1.17), inequality (1.16) follows.

(b) - Let $s = \frac{N}{2}$ and let r_1 be given by (1.17). For any p such that $1 < p < \infty$ it follows that
$$W^{2,s}(\Omega) \subset L^p(\Omega).$$

Choosing $r_0 > 1$, the result follows from Lemma 1.1.

(c) - Let s be such that $\frac{N}{2} < s < N$. In this case we have
$$W^{2,s}(\Omega) \subset C^0(\overline{\Omega}),$$
and
$$W^{2,s}(\Omega) \subset W^{1,r_1}(\Omega),$$
where $r_1 = \frac{Ns}{N-s}$. By using Lemma 1.1 with this r_1 and $r_0 \to \infty$ we get the result.

(d) - Let $N < s$. For some $\alpha > 0$ we have
$$W^{2,s}(\Omega) \subset C^{1,\alpha}(\overline{\Omega}).$$

As a consequence
$$W^{2,s}(\Omega) \subset W^{1,r_1}(\overline{\Omega}) \text{ for any } r_1 > 1.$$

An application of Lemma 1.1 with this arbitrary r_1 and $r_0 = \infty$ concludes the proof.

(d') - Let $N = s$. we have
$$W^{2,s}(\Omega) \subset C^0(\overline{\Omega}).$$
and
$$W^{2,s}(\Omega) \subset W^{1,r_1}(\Omega) \text{ for any } r_1 > 1.$$

By applying Lemma 1.1 with $\dfrac{1}{r} = \dfrac{\tau}{r_1}$ the result follows. Observe that if $\tau > 0$ we can choose any r and if $\tau = 0$ inequality (1.16) is trivial.

By using the same method above we can prove also the following

Corollary 1.2. *Let* $u \in W_0^{1,s}(\Omega) \cap W^{2m,s}(\Omega)$ *where* $1 < s < \infty$, $m \geq 1$ *and* $\tau \in [0, 1]$. *Assume that either*

$$\frac{1}{r} = \frac{1}{s} - \frac{(2m - \tau)}{N} \quad \text{if} \quad s < \frac{N}{2m} \tag{1.18}$$

or

$$\frac{1}{r} = \frac{\tau}{N} \quad \text{if} \quad s = \frac{N}{2m} \tag{1.19}$$

or

$$\frac{1}{r} = \tau(\frac{1}{s} - \frac{1}{N}) \text{ if } \frac{N}{2m} < s < N \tag{1.20}$$

or

$$\forall r > 1, \forall \tau \in [0, 1] \text{ if } N \leq s. \tag{1.21}$$

Then

$$\left\|\frac{u}{\varphi^\tau}\right\|_{L^r} \leq C \|u\|_{W^{2m,s}}, \tag{1.22}$$

where the constant C depends only on τ, s and N.

We conclude this section with some inequalities that will be used in the treatment of the polyharmonic equation (P). Again the proof follows the same pattern of the proof of Corollary 1.1. So for sake of brevity we shall omit it.

Let $m \geq 1$ and $p > 1$. Consider
$$W^{2m,p}_{\#}(\Omega) = \{u \in W^{2m,p}(\Omega) : u = \Delta u = \ldots = \Delta^{m-1} u = 0 \text{ on } \partial\Omega\}.$$

It is not difficult to check that if $1 \leq j \leq m - 1$, then
$$u \in W^{2m,p}_{\#}(\Omega) \Rightarrow \Delta^j u \in W^{2m,p}(\Omega) \cap W^{1,p}_0(\Omega).$$

So in view of Corollary 1.2 we obtain

Corollary 1.3. *Let $u \in W^{2m,p}_{\#}(\Omega)$ and $\tau \in [0,1]$, where $1 < s < \infty$, $m \geq 1$. Assume that $1 \leq j \leq m-1$ and either*

$$\frac{1}{r} = \frac{1}{p} - \frac{(2(m-j)-\tau)}{N} \quad \text{if} \quad p < \frac{N}{2(m-j)} \tag{1.23}$$

or

$$\frac{1}{r} > \frac{\tau}{N} \quad \text{if} \quad p = \frac{N}{2(m-j)} \tag{1.24}$$

or

$$\frac{1}{r} = \tau(\frac{1}{p} - \frac{1}{N}) \text{ if } \frac{N}{2(m-j)} < p < N \tag{1.25}$$

or

$$\forall r > 1, \forall \tau \in [0,1] \text{ if } N \leq p. \tag{1.26}$$

Then for any $1 \leq j \leq m-1$ we have

$$\left\|\frac{\Delta^j u}{\varphi^\tau}\right\|_{L^r} \leq C_1 \left\|\Delta^j u\right\|_{W^{2(m-j),s}} \leq C_2 \left\|u\right\|_{W^{2m,p}}, \tag{1.27}$$

where C_1 depends only on τ, p and N and C_2 also depends on Ω.

2. A priori bounds for positive solutions of system (S)

In this section we shall study existence of a priori bounds of positive solutions of the following class of elliptic systems

$$\begin{cases} -\Delta u = f(x,u,v,Du,Dv) & \text{in } \Omega \subset \mathbf{R}^N, \\ -\Delta v = g(x,u,v,Du,Dv) & \text{in } \Omega \subset \mathbf{R}^N, \\ u = v = 0 & \text{on } \partial\Omega \end{cases} \tag{2.1}$$

where Ω is a smooth bounded domain of $\mathbf{R}^N$ with $N \geq 3$. We shall assume that the nonlinearities f and g satisfy :

(f_1) $f : \overline{\Omega} \times R \times R \times R^N \times R^N \to R$ is continuous.

(f_2)

$$\liminf_{t \to \infty} \frac{f(x,s,t,\xi,\eta)}{t} > \lambda_1 \quad \text{uniformly in } (x,s,\xi,\eta) \in \Omega \times R \times R^N \times R^N.$$

(f_3) There exist $p \geq 1$ and $\sigma \geq 0$ such that

$$|f(x,s,t,\xi,\eta)| \leq C(|t|^p + |s|^{p\sigma}) + 1$$

for any $(x,s,t,\xi,\eta) \in \times R \times R \times R^N \times R^N$.

Similarly we shall assume that :

(g_1) $g : \overline{\Omega} \times R \times R \times R^N \times R^N \to R$ is continuous.

(g_2)
$$\liminf_{s \to \infty} \frac{g(x,s,t,\xi,\eta)}{s} > \lambda_1 \quad \text{uniformly in } (x,t,\xi,\eta) \in \Omega \times R \times R^N \times R^N.$$

(g_3) There exist $q \geq 1$ and $\sigma' \geq 0$ such that
$$|g(x,s,t,\xi,\eta)| \leq C(|s|^q + |t|^{q\sigma'} + 1)$$
for any $(x,s,t,\xi,\eta) \in \overline{\Omega} \times R \times R \times R^N \times R^N$.

Superlinear systems of the type (2.1) with
$$f(x,s,t,\xi,\eta) = f(t) \ , g(x,s,t,\xi,\eta) = g(s)$$
have been treated in Clément, De Figueiredo and Mitidieri [CFM] when Ω is convex and by Peletier-van der Vorst [PvdV] in the case $\Omega = B_R(0)$. In those papers it was observed that the role of the critical exponent that appears when dealing with scalar equations of the form (0.1) has to be replaced by a "*critical hyperbola*". Namely if $f(t) \simeq t^p$ and $g(s) \simeq s^q$ with $p, q > 1$ then, the system
$$-\Delta u = f(v) \qquad -\Delta v = g(u)$$
is *subcritical* if
$$\frac{1}{p+1} + \frac{1}{q+1} > \frac{N-2}{N}, \quad N \geq 3 .$$
The curve
$$\frac{1}{p+1} + \frac{1}{q+1} = \frac{N-2}{N} \qquad (2.2)$$
is called the *critical hyperbola*. The term *subcritical* is justified in view of the results proved in [M$_1$],[CFM][vdV] and [PvdV].

A priori bounds for positive solutions of (2.1) – (2.2) (with further hypotheses on f and g) were established in [CFM] and [PvdV]. The method used in [CFM] depends on the monotonicity of both f and g, while by considering radial solutions, in [PvdV] it was proved that monotonicity is not necessary.

Recently Souto [S] proved a priori bound for positive solutions of (2.1) in the case
$$f(x,s,t,\xi,\eta) = f(x,s,t) \quad , \quad g(x,s,t,\xi,\eta) = g(x,s,t),$$

assuming some growth restrictions on f and g related to the critical exponent $\frac{N+2}{N-2}$. He uses a blow-up technique like in Gidas-Spruck [GS]. This technique for systems is being hampered so far by the lack of adequate Liouville type theorems for positive solutions of the canonical system

$$-\Delta u = v^p \quad , -\Delta v = u^q \quad \text{in} \quad \mathbf{R}^N. \tag{2.3}$$

Some Liouville type theorems for (2.3) have been proved in [M$_1$] for radial positive solutions and in [M$_2$],[S],[FF] for general positive solutions. In particular in [M$_2$] it is proved among other things that (2.3) has no positive solutions if

$$\max\{p+1, q+1\} \geq \frac{N-2}{2}(pq-1).$$

A similar result appears in [S] under a stronger condition on p and q. It is still an open question if there exist positive non-radial solutions of (2.3) under the solely assumption that (p, q) is strictly below the *critical hyperbola* (2.2).

Here we take up the case of general nonlinearities satisfying $(f_1) \ldots (g_3)$ above. However it seems that in general the application of Hardy-Sobolev inequalities apparently precludes the possibility of obtaining optimal results. A similar phenomena appears in the work of Brézis and Turner [BT] when they proved a priori bounds of positive solutions of scalar superlinear elliptic equations. They were able to treat nonlinearities with polynomial growth less than $\frac{N+1}{N-1}$.

In this paper, and for nonlinearities satisfying $(f_1) \ldots (g_3)$, the Brézis-Turner exponent assumption is replaced by conditions that involves two curves in the (p, q) plane. Namely if f and g satisfy $(f_1) \ldots (g_3)$ we will assume that

$$\frac{1}{p+1} + \frac{N-1}{N+1}\frac{1}{q+1} > \frac{N-1}{N+1}$$

and

$$\frac{1}{p+1}\frac{N-1}{N+1} + \frac{1}{q+1} > \frac{N-1}{N+1}.$$

Observe that if $p = q$, these two conditions reduce to the Brézis-Turner assumption

$$p = q < \frac{N+1}{N-1}$$

as expected.

The main result of this paper is the following.

Theorem 2.1. *Let Ω be a smooth bounded domain of R^N with $N \geq 4$. Assume that conditions $(f_1) \ldots (g_3)$ hold with p, q, σ and σ' satisfying :*

$$\frac{1}{p+1} + \frac{N-1}{N+1}\frac{1}{q+1} > \frac{N-1}{N+1}, \tag{2.4}$$

$$\frac{1}{p+1}\frac{N-1}{N+1}+\frac{1}{q+1} > \frac{N-1}{N+1}, \qquad (2.5)$$

and

$$\sigma = \frac{L}{\max(L,K)}, \quad \sigma' = \frac{K}{\max(L,K)}$$

where

$$K = \frac{p}{p+1} - \frac{2}{N} > 0 \text{ and } L = \frac{q}{q+1} - \frac{2}{N} > 0.$$

Let (u,v) be a positive solution of (2.1). Then there exists a constant $C > 0$ independent of (u,v) such that

$$||u||_{L^\infty} \leq C, \ ||v||_{L^\infty} \leq C. \qquad (2.6)$$

Remark 2.1

(i) If $N \geq 4$ and $p, q > 1$ then $K, L > 0$. So if $p > q$ then $\sigma = \frac{L}{K} < 1$ and $\sigma' = 1$. A similar statement holds if $p < q$.

(ii) If $N = 3$ then $K, L > 0$ imply $p, q > 2$ which is not compatible with (2.4). So the case $N = 3$ needs a special treatment and will be studied elsewhere.

(iii) Conditions (2.4) – (2.5) can be written as

$$pq - \frac{2}{N-1}q - \frac{N+1}{N-1} < 0 \quad \text{and} \quad pq - \frac{2}{N-1}p - \frac{N+1}{N-1} < 0, \qquad (2.7)$$

which is equivalent to

$$\min\{p+1, q+1\} > \frac{N-1}{2}(pq-1). \qquad (2.8)$$

Condition (2.7) implies

$$pq - \frac{1}{N-1}p - \frac{1}{N-1}q - \frac{N+1}{N-1} < 0. \qquad (2.9)$$

(iv) A priori bounds for systems using "Hardy-Sobolev" inequalities much like in the spirit of [BT] has been considered by Cosner [C]. However he requires that the growths of the nonlinearities have to be (separately) below the so called Brézis-Turner exponent $\frac{N+1}{N-1}$.

(v) Conditions (2.4) and (2.5) and the assumption $p, q \geq 1$ imply that $p, q < \frac{N+3}{N-1}$. Observe that this bound is strictly less than $\frac{N+2}{N-2}$. This fact will used later.

In the rest of this section if (u,v) is a positive solution of (2.1) we shall write $f = f(x, u, v, Du, Dv)$ and $g = g(x, u, v, Du, Dv)$. As usual we will denote by C a generic constant which is independent of (u,v).

The proof of Theorem 2.1 is organized in three steps. In the first step we prove that f and g are uniformly bounded in $L^1_{loc}(\Omega)$. In the second one we obtain a uniform estimate of the components of the solution (u,v) respectively in $W^{2,\frac{p}{p+1}}(\Omega)$ and $W^{2,\frac{q}{q+1}}(\Omega)$. Finally in the last step we bootstrap.

Proof of Theorem 2.1.

Proof of Step 1.- This makes use of a classical argument, but for sake of completeness we shall repeat it. Let

$$\lambda_1 < l_f < \liminf_{t\to\infty} \frac{f(x,s,t,\xi,\eta)}{t} \tag{2.10}$$

and

$$\lambda_1 < l_g < \liminf_{s\to\infty} \frac{g(x,s,t,\xi,\eta)}{s}. \tag{2.11}$$

FFrom our assumptions $(f_1) - (g_1)$ it follows that there exists $C \geq 0$ such that for all $s, t \geq 0$ and $(x, \xi, \eta) \in \Omega \times R^N \times R^N$ we have

$$f(x,s,t,\xi,\eta) \geq l_f t - C, \tag{2.12}$$

and

$$g(x,s,t,\xi,\eta) \geq l_g s - C. \tag{2.13}$$

Multiplying the equations (2.1) by φ and integrating by parts we get

$$\lambda_1 \int_\Omega u\, \varphi\, dx = \int_\Omega f\, \varphi\, dx \geq l_f \int_\Omega v\, \varphi\, dx - C \int_\Omega \varphi\, dx, \tag{2.14}$$

$$\lambda_1 \int_\Omega v\, \varphi\, dx = \int_\Omega g\, \varphi\, dx \geq l_g \int_\Omega u\, \varphi\, dx - C \int_\Omega \varphi\, dx, \tag{2.15}$$

and then

$$\int_\Omega u\, \varphi\, dx \leq C, \int_\Omega v\, \varphi\, dx \leq C, \tag{2.16}$$

which imply by $(2.14) - (2.15)$ that $|f|$ and $|g|$ are uniformly bounded in $L^1_{loc}(\Omega)$.

Proof of Step 2.- By taking the $L^{\frac{p+1}{p}}$ norm of the first equation of (2.1) we obtain

$$\int_\Omega |\Delta u|^{\frac{p+1}{p}} dx = \int_\Omega |f|^{\frac{p+1}{p}} dx = \int_\Omega |f|^\alpha \varphi^\alpha |f|^{1-\alpha + \frac{1}{p}} \varphi^{-\alpha} dx \tag{2.17}$$

where $0 < \alpha < 1$ is a parameter that will be chosen later. Using Hölder's inequality we obtain

$$\int_\Omega |\Delta u|^{\frac{p+1}{p}} dx = \int_\Omega |f|^{\frac{p+1}{p}} dx \le (\int_\Omega |f| \varphi \, dx)^\alpha (\int_\Omega |f|^{1+\frac{1}{p}\frac{1}{1-\alpha}} \varphi^{-\frac{\alpha}{1-\alpha}} dx)^{1-\alpha}. \tag{2.18}$$

Next by using (f_3) we estimate the second integral of (2.18) with $I_1 + I_2 + C$ where

$$I_1 = \int_\Omega |v|^{p+\frac{1}{1-\alpha}} \varphi^{-\frac{\alpha}{1-\alpha}} dx \tag{2.19}$$

and

$$I_2 = \int_\Omega |u|^{\sigma(p+\frac{1}{1-\alpha})} \varphi^{-\frac{\alpha}{1-\alpha}} dx . \tag{2.20}$$

Now in order to estimate (2.19) and (2.20) we use Corollary 1.1 as follows. For I_1 we choose r and τ such that

$$r = p + \frac{1}{1-\alpha} \quad , \quad r\tau = \frac{\alpha}{1-\alpha} \quad , \quad \frac{1}{r} = \frac{q}{q+1} - \frac{2-\tau}{N} \quad . \tag{2.21}$$

These equations determine α. Indeed

$$\alpha = \frac{N(1-L-pL)}{1+N-pLN} . \tag{2.22}$$

FFrom (2.4) – (2.5) and (iii) of Remark 2.1 it follows that $\alpha > 0$. Using the assumption $L > 0$ we also see that $\alpha < 1$. In order to check that τ determined by (2.21) is such that $0 < \tau < 1$ we observe that from the second equation in (2.21) we get $\tau > 0$. The fact that $\tau < 1$ follows readily from its explicit expression

$$\tau = \frac{N(1-L-pL)}{1+p+N} . \tag{2.23}$$

So using Corollary 1.1 we obtain

$$I_1 \le C \|v\|^r_{W^{2,\frac{q+1}{q}}} , \tag{2.24}$$

and since by (2.22) we have

$$r(1-\alpha) = p(1-\alpha) + 1 = p(1 - \frac{N(1-L-pL)}{1+N-pLN}) + 1 , \tag{2.25}$$

by (2.4) we conclude that

$$\theta_1 := r(1-\alpha) < \frac{q+1}{q} . \tag{2.26}$$

Next we estimate I_2. By choosing $\bar{r}$ and $\bar{\tau}$ such that

$$\sigma(p + \frac{1}{1-\alpha}) = \bar{r} \qquad (2.27)$$

and

$$\frac{\alpha}{1-\alpha} = \bar{r}\,\bar{\tau}\,, \qquad (2.28)$$

from the assumption $\sigma = \frac{L}{\max(L,K)}$ and the expression of α it follows $\bar{r} > 1$.

Now we can see using (2.9) that

$$\frac{\alpha}{1-\alpha} < 1\,. \qquad (2.29)$$

So from (2.27) – (2.28) it follows that

$$0 < \bar{\tau} < 1\,. \qquad (2.30)$$

Next let s be defined by

$$\frac{1}{\bar{r}} = \frac{s}{s+1} - \frac{2-\bar{\tau}}{N}\,. \qquad (2.31)$$

Rewriting (2.31) we get

$$\frac{s}{s+1} = \frac{1}{\bar{r}} + \frac{2-\bar{\tau}}{N} = \frac{1}{\sigma}(\frac{1}{r} - \frac{\tau}{N}) + \frac{2}{N} = \frac{1}{\sigma}L + \frac{2}{N} = \max(L,K) + \frac{2}{N}\,, \qquad (2.32)$$

so we infer that $s = \max(p,q)$.

We are now in position to apply again Corollary 1.1. We obtain

$$I_2 \leq C\|u\|_{W^{2,\frac{s+1}{s}}}^{\bar{r}}\,, \qquad (2.33)$$

and by (2.26)

$$\bar{r}(1-\alpha) = \sigma(p(1-\alpha)+1) < \sigma(\frac{q+1}{q})\,. \qquad (2.34)$$

Since

$$\sigma(\frac{q+1}{q}) \leq \frac{p+1}{p}\,,$$

we finally conclude that

$$\theta_2 := \bar{r}(1-\alpha) < \frac{p+1}{p}\,. \qquad (2.35)$$

Since

$$\frac{s+1}{s} = \min(\frac{p+1}{p}, \frac{q+1}{q})\,,$$

it follows from step 1 and the relations (2.24), (2.26), (2.33) and (2.35) that

$$\int_\Omega |\Delta u|^{\frac{p+1}{p}} dx \leq C(||v||_{W^{2,\frac{q+1}{q}}}^{\theta_1} + ||u||_{W^{2,\frac{p+1}{p}}}^{\theta_2} + 1) . \tag{2.36}$$

Observing that in this context the roles of p and q are symmetric, following a similar reasoning we also find that

$$\int_\Omega |\Delta v|^{\frac{q+1}{q}} dx \leq C(||u||_{W^{2,\frac{p+1}{p}}}^{\theta_3} + ||v||_{W^{2,\frac{q+1}{q}}}^{\theta_4} + 1) , \tag{2.37}$$

where $\theta_3 < \frac{p+1}{p}$ and $\theta_4 < \frac{q+1}{q}$.

Using the inequality

$$||u||_{W^{2,t}} \leq C ||\Delta u||_{L^t} , \tag{2.38}$$

which holds for all $u \in W_0^{1,t}(\Omega) \cap W^{2,t}(\Omega)$, with $t > 1$, we get

$$||u||_{W^{2,\frac{p+1}{p}}}^{\frac{p+1}{p}} \leq C(||u||_{W^{2,\frac{p+1}{p}}}^{\frac{p+1}{p}\gamma_1} + ||v||_{W^{2,\frac{q+1}{q}}}^{\frac{q+1}{q}\gamma_2} + 1) , \tag{2.39}$$

$$||v||_{W^{2,\frac{q+1}{q}}}^{\frac{q+1}{q}} \leq C(||u||_{W^{2,\frac{p+1}{p}}}^{\frac{p+1}{p}\gamma_3} + ||v||_{W^{2,\frac{q+1}{q}}}^{\frac{q+1}{q}\gamma_4} + 1) , \tag{2.40}$$

where $0 < \gamma_i < 1$, $i = 1...4$. Thus we conclude that

$$||u||_{W^{2,\frac{p+1}{p}}} \leq C , \quad ||v||_{W^{2,\frac{q+1}{q}}} \leq C . \tag{2.41}$$

Proof of step 3.- In this step we bootstrap in order to get bounds in stronger norms. Since we are away from the critical hyperbola we have more room than in [CFM]. This allow us to use more rough inequalities than (f_3) and (g_3) in order to simplify the various steps in the bootstrap process. From our assumptions on K and L it follows that $\sigma, \sigma' \leq 1$. Whence from (f_3) and (g_3) we get

$$|f| \leq C(|u|^p + |v|^p + 1) \tag{2.42}$$

$$|g| \leq C(|u|^q + |v|^q + 1) . \tag{2.43}$$

As a consequence, we may assume without loss of generality that $p \geq q$ and

$$|f| , |g| \leq C(|u|^p + |v|^p + 1) . \tag{2.44}$$

FFrom step 2 it follows that

$$||u||_{L^k} \leq C , \quad ||v||_{L^t} \leq C . \tag{2.45}$$

87

Since $p \geq q$ we have $\frac{1}{K} \geq \frac{1}{L}$. So we start our bootstrap process with

$$\|u\|_{L^{\frac{1}{K}}}, \|v\|_{L^{\frac{1}{K}}} \leq C$$

and use the growth conditions on f and g given by (2.44). Now the procedure is like in the scalar case and it works because in view of (v) - Remark 2.1 we have that the growth p of f is strictly less than $\frac{N+2}{N-2}$ and hence $\frac{1}{K} \geq \frac{2N}{N-2}$. This concludes the proof of the theorem.

3. A priori bounds for positive solutions of the polyharmonic problem (P)

Let $m > 1$ and let Ω be a smooth bounded domain of $\mathbf{R}^N$ with $N > 2m$. In this section we shall prove a priori bounds for a polyharmonic problem of the type

$$\begin{cases} (-\Delta)^m u = h(x, u, \Delta u, \ldots, \Delta^{m-1} u) & \text{in } \Omega \subset \mathbf{R}^N, \\ u = \Delta u = \ldots = \Delta^{m-1} u = 0 & \text{on } \partial \Omega. \end{cases} \quad (3.1)$$

The assumptions on the function h are the following :

(h_1) $h : \overline{\Omega} \times R^m \to R$ is continuous,

(h_2)
$$\liminf_{s \to \infty} \frac{h(x, s, t_1, \ldots t_{m-1})}{s} > \lambda_1{}^m$$

uniformly in $(x, t_1, \ldots t_{m-1}) \in \Omega \times R^{m-1}$,

(h_3) there exist $C > 0$ and $p > 1$ such that

$$|h(x, s, t_1, \ldots t_{m-1})| \leq C(1 + |s|^p)$$

for every $(x, s, t_1, \ldots t_{m-1}) \in \Omega \times R^m$.

Theorem 3.1. *Assume (h_1)-(h_3) together with*

$$\frac{2m-1}{N-2m+1} < p < \frac{N+1}{N-2m+1}. \quad (3.2)$$

Let u be a positive solution of (3.1). Then there exists a constant $C > 0$ independent of u such that

$$\|u\|_{L^\infty} \leq C. \quad (3.3)$$

Proof. We shall organize the proof in three steps.

Step 1.- As in step 1 of the proof of Theorem 2.1 we obtain

$$\int_\Omega |h|\varphi \, dx \leq C . \tag{3.4}$$

That is, h is uniformly bounded in $L^1_{loc}(\Omega)$.

Step 2.- By taking the $L^{\frac{p+1}{p}}(\Omega)$ norm of both sides of (3.1) we obtain

$$\int_\Omega |\Delta^m u|^{\frac{p+1}{p}} dx = \int_\Omega |h|^{\frac{p+1}{p}} dx = \int_\Omega |h|^\alpha \varphi^\alpha |h|^{1-\alpha+\frac{1}{p}} \varphi^{-\alpha} dx \tag{3.5}$$

where $0 < \alpha < 1$ is a parameter to be determined later.

Using Hölder's inequality and elliptic regularity we obtain

$$||u||_{W^{2m,\frac{p+1}{p}}}^{\frac{p+1}{p}} \leq C (\int_\Omega |h|\varphi \, dx)^\alpha (\int_\Omega |h|^{1+\frac{1}{p}\frac{1}{1-\alpha}} \varphi^{-\frac{\alpha}{1-\alpha}} dx)^{1-\alpha} . \tag{3.6}$$

In view of (h_3), the second integral in the right hand side is estimated by $I + C$, where

$$I = \int_\Omega |u|^{p+\frac{1}{1-\alpha}} \varphi^{-\frac{\alpha}{1-\alpha}} dx .$$

Define

$$L = \frac{p}{p+1} - \frac{2m}{N} . \tag{3.7}$$

Observe that by (3.2) we have

$$1 + NL > 0 .$$

Viewing the use of Corollary 1.2 (i) we set

$$p + \frac{1}{1-\alpha} = r, \quad \frac{\alpha}{1-\alpha} = r\tau, \quad \frac{1}{r} = \frac{p}{p+1} - \frac{2m-\tau}{N} . \tag{3.8}$$

This set of equations with L given in (3.7) is exactly like (2.21). Hence they determine α, τ and r :

$$\alpha = \frac{N - NL - NLp}{1 + N - pNL}, \quad \tau = \frac{N - NL - NLp}{1 + N + p}, \quad r = \frac{1 + N + p}{1 + NL} . \tag{3.9}$$

First we claim that $1 + N - NLp > 0$. Obviously if $L \leq 0$ this last condition is satisfied. Let us suppose that $L > 0$. In view of the conditions on p set in (3.2), it suffices to prove that $1 + N > \frac{NL(N+1)}{(N-2m+1)}$, which is easily seen to be true. On the other hand, it is immediate to verify that $1 - L - Lp > 0$ is equivalent to $p < \frac{N+2m}{N-2m}$.

89

The bound on p set in (3.2) is smaller that this value; so $\alpha, \tau > 0$. On the other hand, it follows readily that $\alpha, \tau < 1$ hence from (3.8) we conclude that $r > 1$.

Now in order to estimate I, we use Corollary (1.2)-(i) as follows

$$I \leq C ||u||_{W^{2m, \frac{p+1}{p}}}^r .$$

Observe that $\theta := r(1-\alpha) = p(1-\alpha) + 1$ in view of (3.7) and it follows from (3.2) that $\theta < \frac{(p+1)}{p}$. Let us emphasize that, since we do not need this last inequality to be strict, the best value of p that can handled by this method is the one set in (3.2).

Proof of step 3.- This involves a bootstrap argument like in the proof of Theorem (2.1) so we shall omit it.

This completes the proof of the Theorem.

Acknowledgments. Djairo G. de Figueiredo was partially supported by CNPq (Brazil). Enzo Mitidieri acknowledges the support of MURST (40% and 60%).

REFERENCES

[BT] H. Brézis & R. Turner, *On a Class of Superlinear Elliptic Problems*, Comm. Partial Diff. Equations, 2, (1977) 601-614.

[CFM] Ph. Clément, D.G. de Figueiredo & E. Mitidieri, *Positive Solutions of Semilinear Elliptic Systems*, Comm. Partial Diff. Equations,Vol. 17 (5&6), No.,(1993), 923-940.

[C] C. Cosner, *Positive Solutions for Superlinear Elliptic Systems without Variational Structure*, Nonlinear Analysis T.M.A.,Vol. 8, 12, (1984), 1427-1436.

[FF] D.G. de Figueiredo & P. Felmer, A Liouville type theorem for elliptic systems, Ann. Scuola Normale di Pisa, Serie IV, XXI., (1994), 387-397.

[FLN] D.G. de Figueiredo, R. D. Nusbaum & P.L. Lions, *A Priori Estimates and Existence of Positive Solutions of Semilinear Elliptic Equations*, J. Math. Pures et Appl., Vol. 61, (1982), 41-63.

[GS] B. Gidas & J. Spruck, *A Priori Bounds for Positive Solutions of Nonlinear Elliptic Equations*, Comm. Partial Diff. Equations, 6, (1981), 883-901.

[HLP] G.H. Hardy, J.E. Littlewood & G. Pólya, *Inequalities*, Cambridge University Press (1967).

[K] O. Kavian, *Inegalité de Hardy-Sobolev et Application*, Thèse de Doctorat de 3eme cycle, Université de Paris VI, (1978).

[LM] J.L. Lions & E. Magenes, Non-homogeneous Boundary Value Problems and Applications, Vol. 1, Springer Verlag (1972).

[MA] V. G. Mazja, *Sobolev Spaces*, Springer-Verlag, (1985).

[M_1] E. Mitidieri, *A Rellich Type Identity and Applications*, Comm. Partial Diff. Equations, 18 (1&2), (1993), 125-151.

[M_2] E. Mitidieri, *Non-existence of Positive Solutions of Semilinear Elliptic Systems* in R^N, Quaderno Matematico N. 285, (1992) (to appear in Differential & Integral Equations).

[PL] P. L. Lions, *On the existence of positive solutions of semilinear elliptic equations*, SIAM Review, 24, (1982), 441-467.

[PS] P. Pucci & J. Serrin, *A General Variational Identity*, Indiana Univ. Math. J., 35, (1986), 681-703.

[PvdV] L.A. Peletier & R. C. A. M. van der Vorst, *Existence and Non-existence of Positive Solutions of Nonlinear Elliptic Systems and the Biharmonic Equation*, Differential & Integral Equations, 54, (1991).

[S] M.A. Souto, *Sobre a existencia de solucoes positivas de sistemas cooperativos nao lineares*, Tese de doutorado, UNICAMP, (1992).

[vdV] R. C. A. M. van der Vorst, *Variational Identities and Applications to Differential Systems*, Arch. Rational Mech. Anal., 116, (1991) 375-398.

[Z] W.P. Ziemer, *Weakly Differentiable Functions*, Springer-Verlag (1989).

ADDRESS

Philippe Clément, Department of Technical Mathematics and Informatics. University of Technology, Delft. The Netherlands

Djairo Guedes de Figueiredo, IMECC - UNICAMP. Caixa Postal 6065 - 13081-970 Campinas, S.P.Brazil

Enzo Mitidieri, Dipartimento di Scienze Matematiche. Università degli Studi di Trieste Piazzale Europa 1 34100. Trieste, Italy

B DACOROGNA
On the minimisation of non quasiconvex integrals of the calculus of variations

1. Introduction and relaxation theorem

Let

$$F(u) = \int_\Omega f(\nabla u(x))dx \tag{1.1}$$

where $\Omega \subset \mathbb{R}^n$ is a bounded open set, $u: \Omega \to \mathbb{R}^m$ and thus $\nabla u = \left(\dfrac{\partial u^\alpha}{\partial x_i}\right)_{1 \leq i \leq n, 1 \leq \alpha \leq m} \in \mathbb{R}^{nm}$ and $f: \mathbb{R}^{nm} \to \mathbb{R}$ is a lower semicontinuous function.

Finally let $u_0 \in W^{1,\infty}(\Omega; \mathbb{R}^m)$ (where $W^{1,\infty}$ denotes the usual Sobolev space) and

(P) $\quad \inf\{F(u): u \in u_0 + W_0^{1,\infty}(\Omega; \mathbb{R}^m)\}$.

To show the existence of minima for such problems in $W^{1,p}$ ($1 < p < \infty$), one has, in general to make a coercitivity hypothesis $\left(\text{i.e. } f(\xi) \geq \alpha |\xi|^p + \beta \text{ for } \alpha > 0\right)$ and a hypothesis of semicontinuity of F with respect to weak convergence in $W^{1,p}$. A necessary and sufficient condition to obtain such a property is the notion of quasiconvexity introduced by Morrey [1].

Definition

A lower semicontinuous function $f: \mathbb{R}^{nm} \to \mathbb{R}$, is said to be <u>quasiconvex</u> if

$$\int_\Omega f(\xi + \nabla\varphi(x))dx \geq f(\xi) \cdot \text{mes}\,\Omega \tag{1.2}$$

for every $\xi \in \mathbb{R}^{nm}$ and for every $\varphi \in W_0^{1,\infty}(\Omega; \mathbb{R}^m)$.

Remarks:

(i) Jensen inequality ensures, trivially, that every convex function is quasiconvex. If $m = 1$ or if $n = 1$, these are in fact the only quasiconvex functions.

(ii) However as soon as $m, n > 1$, there exist quasiconvex functions that are not convex. A typical example of a quasiconvex function when $m = n = 2$ is

$$f(\xi) = g(\xi, \det \xi)$$

with $g: \mathbb{R}^{2 \times 2} \times \mathbb{R} \to \mathbb{R}$ convex (in particular $f(\xi) = |\det \xi|$). Or if

$$f(\xi) = |\xi|^2 \left(|\xi|^2 - 2 \det \xi \right)$$

we have that f is quasiconvex but not convex. For more details one can refer to Dacorogna [2] or to Alibert-Dacorogna [1].

In this article we will discuss the existence of minima for (P) when the function f is not quasiconvex. In general problem (P) will not have a solution as soon as $n > 1$.

Remark

The case $n = 1$ is peculiar and very simple. A necessary and sufficient condition for the existence of minima is given, for example, in Dacorogna [2]. In particular if f is coercive (P) always has a solution independently of the convexity of f, this is obviously not the case as soon as $n > 1$ or even if $n = 1$ and f depends explicitly on u and not only on u'.

In the general case $n, m \geq 1$ and even if (P) has no solution one can always associate a relaxed problem and one then has

Relaxation theorem

Let Ω, f and u_o as above. Let Qf be the quasiconvex envelope of f i.e.

$$Qf(\xi) = \sup\{\varphi(\xi): \varphi \text{ quasiconvex and } \varphi \leq f\} \qquad (1.3)$$

and

(QP) $\quad \inf\left\{ \overline{F}(u) = \int_\Omega Qf(\nabla u(x)) dx : u \in u_o + W_o^{1,\infty}(\Omega; \mathbb{R}^m) \right\}$

then $\quad \forall \bar{u} \in u_o + W_0^{1,\infty}(\Omega; \mathbb{R}^m), \exists u_v \in u_o + W_0^{1,\infty}(\Omega; \mathbb{R}^m)$

$$u_v \xrightarrow{*w^{1,\infty}} \bar{u}$$

$$F(u_v) = \int_\Omega f(\nabla u_v(x))\, dx \to \bar{F}(\bar{u}) = \int_\Omega Qf(\nabla \bar{u}(x))dx$$

hence in particular

$$\inf(P) = \inf(QP). \tag{1.4}$$

Remarks

(i) By $u_v \xrightarrow{*w^{1,\infty}} \bar{u}$, we mean that u_v converge weak * in $W^{1,\infty}$ to $\bar{u}$.

(ii) The above theorem has been established by Dacorogna [1],[2], generalising to the vectorial case (i.e. $m > 1$) the results of L.-C. Young, Mac Shane, Ekeland... (for more details cf. Dacorogna [2]).

(iii) If, in addition, f satisfies a coercitivity hypothesis, the problem (QP) has a solution in $W^{1,p}$.

(iv) Obviously the theorem does not say anything about the existence or the non existence of solutions for problem (P) but only about the minimising sequences. However it ensures that if (P) has a solution $\bar{u} \in u_o + W_0^{1,\infty}(\Omega; \mathbb{R}^m)$ then from (1.4) and from the fact that

$$Qf(\xi) \le f(\xi) \tag{1.5}$$

for every $\xi \in \mathbb{R}^{nm}$, $\bar{u}$ is also a solution of (QP) and

$$Qf(\nabla \bar{u}) = f(\nabla \bar{u}) \text{ a.e..} \tag{1.6}$$

(Note that in the case $m = 1$, the equation (1.6) is an equation of <u>Hamilton-Jacobi</u> type).

(v) The actual computation of the quasiconvex envelope defined by (1.3) is, in general, difficult (cf. below). Note that if $n = 1$ or if $m = 1$ then the quasiconvex envelope is nothing else than the <u>convexe envelope</u> of f i.e.

$$Cf(\xi) = \sup\{\varphi(\xi): \varphi \text{ convex and } \varphi \leq f\}. \qquad (1.7)$$

In the remaining part of the article we will restrict ourselves to the case where the boundary datum is <u>linear</u> i.e.

$$u_o(x) = \xi_o x \qquad (1.8)$$

where $\xi_o \in \mathbb{R}^{nm}$. The problem (QP) has therefore as a trivial solution u_o itself. We are then going to study the existence of solutions for problem (P). We will follow Dacorogna-Marcellini [1] and we refer to this article for a larger bibliography on this problem which has been studied by many authors.

2. SUFFICIENT CONDITION FOR THE EXISTENCE OF MINIMA

We introduce first some notations

$$\xi = \begin{pmatrix} \xi_1^1 & \cdots & \xi_n^1 \\ \vdots & & \vdots \\ \xi_1^m & \cdots & \xi_n^m \end{pmatrix} = \begin{pmatrix} \xi^1 \\ \vdots \\ \xi^m \end{pmatrix} \in \mathbb{R}^{nm}.$$

Let $\Omega \subset \mathbb{R}^n$ be bounded and open, $f: \mathbb{R}^{nm} \to \mathbb{R}$ a lower semicontinuous function and

$$K = \{\xi \in \mathbb{R}^{nm}: Qf(\xi) < f(\xi)\}. \qquad (2.1)$$

We then let $\xi_o \in K$ (recall that the boundary datum is $u_o(x) = \xi_o x$). By abuse of notations, we will identify K with its connected component which contains ξ_o. We then introduce a manifold L_o whose dimension is n and is defined by

$$L_o = \{\xi = (\xi^\alpha)_{1 \leq \alpha \leq m} \in \mathbb{R}^{nm}: \xi^\alpha = \xi_o^\alpha + \mu^\alpha(\xi_o^m - \xi^m), \alpha = 1, 2, \ldots, m-1\} \qquad (2.2)$$

where $\mu^\alpha \in \mathbb{R}$ (if $m = 1$ we just let $L_o = \mathbb{R}^n$). We then have the following theorem

Theorem 2.1

Let Ω, f, u_o and (P) as above. If $\xi_o \in K$ and if

(i) $K \cap L_o$ is bounded and convex

(ii) Qf is quasiaffine on $K \cap L_o$

then (P) has a solution.

Remarks :

(i) In the case $n = m = 2$, by Qf quasiaffine on $K \cap L_o$ we mean that there exists $\lambda \in \mathbb{R}^{2 \times 2}$ and $\mu, v \in \mathbb{R}$ such that

$$Qf(\xi) = <\lambda; \xi> + \mu \det \xi + v = \sum_{i,\alpha=1}^{2} \lambda_i^\alpha \xi_i^\alpha + \mu \det \xi + v \qquad (2.3)$$

for every $\xi \in K \cap L_o$.

(ii) In the general case $n, m \geq 1$, by Qf quasiaffine we mean that Qf is the sum of an affine function and a linear combination of minors of the matrix ξ.

(iii) The hypothesis $K \cap L_o$ bounded can be considerably weakened, it suffices to assume that $K \cap L_o$ is bounded in certain directions. More precisely

$$K \cap L_o = \left\{ \xi = (\xi^\alpha)_{1 \leq \alpha \leq m} \in L_o \subset \mathbb{R}^{nm} : \left(<b_j; \xi^m>_{\mathbb{R}^n} \right)_{1 \leq j \leq \ell} \text{ bounded in } \mathbb{R}^\ell \right\} (2.4)$$

where ℓ is an integer between 1 and n and $b_1, \ldots, b_\ell$ are linearly independent vectors ($<.;.>_{\mathbb{R}^n}$ denoting the scalar product).

(iv) Finally note that if $\xi_o \notin K$, then problem (P) has u_o itself as a trivial solution.

Example

Consider the case $m = n = 2$ and

$$f(\xi) = g(\xi^1) + h(\det \xi) = g(\xi_1^1, \xi_2^1) + h(\det \xi) \qquad (2.5)$$

where $g: \mathbb{R}^2 \to \mathbb{R}$ is convex (in particular $g \equiv o$) and $h: \mathbb{R} \to \mathbb{R}$ is nonconvex. It is easy to see (cf. Dacorogna-Marcellini [1]) that

$$Qf(\xi) = g(\xi^1) + Ch(\det \xi) \tag{2.6}$$

where Ch is the convex envelope of h. We also trivially have

$$K = \{\xi \in \mathbb{R}^{2\times 2}: Qf(\xi) < f(\xi)\} = \{\xi \in \mathbb{R}^{2\times 2}: Ch(\det \xi) < h(\det \xi)\}. \tag{2.7}$$

As $h: \mathbb{R} \to \mathbb{R}$, we will assume that the connected component of K which contains ξ_o is of the form

$$K = \{\xi \in \mathbb{R}^{2\times 2}: \det \xi \in (\alpha, \beta)\} \tag{2.8}$$

for $\alpha, \beta \in \mathbb{R}$ (note that the connected component of K which contains ξ_o is always of this form, with however the possibility that $\alpha = -\infty$ or $\beta = +\infty$). We then choose in (2.2) $\mu = 0$, we therefore obtain

$$L_o = \{\xi \in \mathbb{R}^{2\times 2}: \xi^1 = (\xi_o)^1\}. \tag{2.9}$$

If we then take in (2.4)

$$b = \left(-(\xi_o)^1_2, (\xi_o)^1_1\right) \in \mathbb{R}^2 \tag{2.10}$$

we indeed have that $K \cap L_o$ is bounded and convex since

$$K \cap L_o = \{\xi \in L_o \subset \mathbb{R}^{2\times 2}: <b; \xi^2> = \det \xi \in (\alpha, \beta)\}. \tag{2.11}$$

Finally observe that on $K \cap L_o$ we indeed have that Qf is affine. Since the first component ξ^1 is fixed $\left(=(\xi_o)^1\right)$ and where $Ch \neq h$ we have that Ch is affine (since $h: \mathbb{R} \to \mathbb{R}$). Theorem 2.1 then applies and we get

Corollary 2.2

Let

$$f(\xi) = g(\xi^1) + h(\det \xi)$$

and $\xi_o \in \mathbb{R}^{2\times 2}$. If g is convex and

$$\lim_{|\delta|\to\infty}\frac{h(\delta)}{|\delta|}=\infty \qquad (2.12)$$

then (P) has a solution.

Remark

The hypothesis (2.12) implies that in (2.8) $\alpha \neq -\infty$ and $\beta \neq \infty$.

We next mention a second example which has been intensively studied by Kohn-Strang [1] and which finds its origins in problems of optimal design. Let

$$f(\xi)=\begin{cases} 1+|\xi|^2 & \text{if } \xi \neq 0 \\ 0 & \text{if } \xi = 0 \end{cases}. \qquad (2.13)$$

One can then show

Theorem 2.3

Let $n = m = 2$ then (P) has a solution if and only if one of the following two conditions hold

(i) $\xi_o = 0$ or $|\xi_o|^2 + 2|\det \xi_o| \geq 1$

(ii) $\det \xi_o \neq 0$.

Remarks

(i) The above result, including its generalisation to $m \geq 2$, is proved in Dacorogna-Marcellini [1]. Note that part (i) of the theorem corresponds to the case where $Qf(\xi_o) - f(\xi_o)$. Indeed Kohn Strang have shown that

$$Qf(\xi)=\begin{cases} f(\xi)=1+|\xi|^2 & \text{if } |\xi|^2+2|\det \xi| \geq 1 \\ 2\left(|\xi|^2+2|\det \xi|\right)^{1/2}-2|\det \xi| & \text{if } |\xi|^2+2|\det \xi| \leq 1 \end{cases}. \qquad (2.14)$$

(ii) For the necessary part of the theorem, cf. next section.

(iii) Theorem 2.3 does not enter in the framework of Theorem 2.1. The similarity comes from the fact that if we choose

$$L = \{\xi \in \mathbb{R}^{2\times 2}: \xi = \xi^t, \text{ i.e. } \xi_2^1 = \xi_1^2 \text{ and } \det \xi > 0\} \quad (2.15)$$

and if $\det \xi_o > 0$ and trace $(\xi_o) \in (0,1)$ then

$$K \cap L = \{\xi \in \mathbb{R}^{2\times 2}: \xi_2^1 = \xi_1^2, \det \xi > 0 \text{ and trace } (\xi) = \xi_1^1 + \xi_2^2 \in (0,1)\} \quad (2.16)$$

Moreover on $K \cap L$

$$Qf(\xi) = 2(\xi_1^1 + \xi_2^2) - 2\det \xi \quad (2.17)$$

i.e. Qf is quasiaffine on $K \cap L$.

Before discussing the generalisation of these examples, we examine a necessary condition.

3. NECESSARY CONDITION FOR THE EXISTENCE OF MINIMA

Let us recall that the problem under consideration is

(P) $\quad \inf\left\{F(u) = \int_\Omega f(\nabla u(x))dx : u \in u_o + W_o^{1,\infty}(\Omega; \mathbb{R}^m)\right\}$

where $u_o(x) = \xi_o x$ with $\xi_o \in \mathbb{R}^{nm}$. We assume that

$$\xi_o \in K = \{\xi \in \mathbb{R}^{nm}: Qf(\xi) < f(\xi)\}. \quad (3.1)$$

To express the necessary condition we must introduce the following definition

Definition

Let $h: \mathbb{R}^{nm} \to \mathbb{R}$ be convex. We say that h is strictly convex in $\xi_o = ((\xi_o)_i^\alpha)_{1\le i\le n, 1\le \alpha\le m} = (\xi_o^\alpha)_{1\le \alpha\le m} \in \mathbb{R}^{nm}$ in m directions if there exists $\lambda = (\lambda^\alpha)_{1\le \alpha \le m} \in \mathbb{R}^{nm}$ such that

$$\lambda^\alpha \neq 0 \text{ and } <\lambda^\alpha; \xi^\alpha - \xi_o^\alpha>_{\mathbb{R}^n} = 0 \quad \forall \alpha = 1, 2, \ldots, m \qquad (3.2)$$

for every $\xi = (\xi^\alpha)_{1 \leq \alpha \leq m}$ which satisfies

$$\frac{h(\xi) + h(\xi_o)}{2} = \frac{h(\xi + \xi_o)}{2}. \qquad (3.3)$$

Remarks

(i) If $m = 1$ the above definition simply means that h is not affine in the neighbourhood of ξ_o, i.e. there exists at least one direction of strict convexity.

(ii) Note also that if h is strictly convex in the neighbourhood of ξ_o then it has at least m directions of strict convexity (in fact nm directions) since (3.3) holds only if $\xi = \xi_o$.

We can then state the following theorem

Theorem 3.1

Let $\xi_o \in \mathbb{R}^{nm}$ be such that

(i) $Cf(\xi_o) = Qf(\xi_o) < f(\xi_o)$

(ii) Cf is strictly convex in ξ_o in at least m directions

then (P) has no solution.

Remark

The theorem has been established by Dacorogna-Marcellini [1]. The weak point of this theorem is that we assume that $Cf(\xi_o) = Qf(\xi_o)$, which is very restrictive. Part (ii) of the theorem is however more interesting since it gives a necessary condition for the existence of a minimum; namely Cf should be affine in at least $nm - (m-1)$ directions (cf. next section for the comparison with the sufficient condition).

We can then deduce two corolaries.

Corollary 3.2

Let $m = 1$, $\xi_o \in \mathbb{R}^n$ and $Cf(\xi_o) < f(\xi_o)$. If (P) has a solution then Cf is affine in the neighbourhood of ξ_o.

Remarks :

(i) The corollary is just a restatement of the theorem since in the case $m = 1$ we always have $Cf = Qf$.

(ii) Furthermore it is interesting to compare this result with the sufficient condition. We see that if, in addition, the connected component of $K = \{Cf(\xi) < f(\xi)\}$ which contains ξ_o is bounded, then the condition is also sufficient (cf. Dacorogna-Marcellini [1] for more details).

Corollary 3.3

Let $\xi_o \in \mathbb{R}^{nm}$ and

$$f(\xi) = g(|\xi|) \tag{3.4}$$

where g is extended to be even to the whole of $\mathbb{R}$. If $n \geq 2$ and if

(i) $Cg(|\xi_o|) < g(|\xi_o|)$

(ii) Cg is strictly increasing in the neghbourhood of $|\xi_o|$

(iii) $rank\{\xi_o\} \leq 1$

then (P) has no solution.

Remark

In the case where

$$g(t) = \begin{cases} 1+t^2 & \text{si } t \neq 0 \\ 0 & \text{si } t = 0 \end{cases}$$

(cf. Theorem 2.3) and $m = n = 2$ we see that the condition $\det \xi_o \neq 0$ (and to avoid the trivial case $Cg(|\xi_o|) \leq Qf(\xi_o) < f(\xi_o)$) is a necessary and sufficient condition for the existence of solutions.

4. FINAL REMARKS

We will now compare the necessary and sufficient conditions of the preceding sections (theorems 2.1 and 3.1). First the hypothesis $K \cap L_o$ <u>bounded</u> and <u>convex</u> appears only as a mean for solving a system of equations of de Hamilton-Jacobi type (cf.(1.6) and Dacorogna-Marcellini [1] pfor more details).

We now turn to the hypothesis $Cf(\xi_o) = Qf(\xi_o)$ in Theorem 3.1. Obviously it is too restrictive and the necessary condition is then very far from the sufficient condition (only in certain instances this is not the case, cf. corollaries 3.2 and 3.3).

However parts (ii) of theorems 2.1 and 3.1 are closer than they appear. The necessary condition requires that Qf be affine in at least $nm - (m-1)$ directions, while the sufficient condition requires the (quasi) affinity of Qf on $K \cap L_o$ which is contained on the manifold L_o of dimension $nm - n(m-1) = n$. The difference between the two seems very important, however this is only an appearence since (cf. Dacorogna-Marcellini [1]).

Proposition 4.1

Let

$$L = \left\{ \xi = (\xi^\alpha)_{1 \leq \alpha \leq m} \in \mathbb{R}^{nm} : \sum_{i=1}^{n} a_i^\alpha (\xi_i^\alpha + \mu^\alpha \xi_i^m) = c^\alpha ; \alpha = 1, 2, \ldots, m-1 \right\}$$

where $c^\alpha, \mu^\alpha \in \mathbb{R}$, $\lambda^\alpha = (a_i^\alpha)_{1 \le i \le n}$ are non zero vectors of $\mathbb{R}^n$ (if m=1, we let $L = \mathbb{R}^n$). Finally let $\Omega \subset \mathbb{R}^n$, $u_o(x) = \xi_o x$ with $\xi_o \in L$ and $u \in u_o + W_o^{1,\infty}(\Omega; \mathbb{R}^m)$ then

$$\nabla u(x) \in L \text{ a.e.} \Leftrightarrow \nabla u(x) \in L_o \text{ a.e.}.$$

Remarks :

(i) Observe, this time, that the dimension of L is $nm - (m-1)$, which is the same as the one appearing in the necessary condition. Furthermore $L_o \subset L$, trivially since $\xi_o \in L$.

(ii) The proposition shows that if we fix $(m-1)$ conditions (i.e. $\nabla u(x) \in L$) and the boundary condition, then automatically $n(m-1)$ conditions are fixed (i.e. $\nabla u(x) \in L_o$).

To conclude these remarks we dicuss briefly the problem of the actual computation of the quasiconvex envelope Qf. This is a delicate problem and we refer to Dacorogna [2] for some general formulas and examples; we should also mention the recent results of Buttazzo-Dacorogna-Gangbo [1] and Dacorogna-Koshigoe [1] which allows certain simplifications in the computation of Qf when the function has certain symetries).

BIBLIOGRAPHY

- J.J. Alibert - B. Dacorogna [1] : An example of a quasiconvex function that is not polyconvex in two dimensions; Arch. Rational Mech. Anal. 117 (1992), 155-166.

- G. Buttazzo - B. Dacorogna - W. Gangbo [1] : On the envelopes of functions depending on singular values of matrices; Boll. U.M.I. 8-B (1994), 17-35.

- B. Dacorogna [1] : Quasiconvexity and relaxation of nonconvex problems in the calculus of variations; J. Funct. Anal. 46 (1982), 102-118.

- B. Dacorogna [2] : "Direct methods in the calculus of variations", Springer, Berlin (1989).

- B. Dacorogna - H. Koshigoe [1] : On the different notions of convexity for rotationally invariant functions; Annales Fac. Sciences de Toulouse (1993), 163-184.

- B. Dacorogna - P. Marcellini [1] : Existence of minimizers for non quasiconvex integrals; à paraître dans Arch. Rational Mech. Anal.

- R.V. Kohn - G. Strang [1] : Optimal design and relaxation of variational problems I, II, III; Commun. Pure Appl. Math. 39 (1986), 113-137; 139-182; 353-377.

- C.B. Morrey [1] : "Multiple integrals in the calculus of variations"; Springer, Berlin (1966).

Bernard DACOROGNA, Ecole Polytechnique Fédérale de Lausanne, Department of Mathematics - CH 1015 LAUSANNE

B DEHMAN
Local solvability for complex quasi-linear equations

The property of local solvability for complex differential operators have been investigated by many authors. Besides the fundamental result of Cauchy-Kowalevsky, we can mention essentialy the works of Nirenberg-Trêves [8] and Hörmander (see for example [5]) for the linear case.

Concering non linear equations, only few results are available. In this context we have the paper of [9], the recent work of Trêves [10] and that of the author [2] in wich a semi-linear case is treated. We can alsow mention the work of Gramchev-Popivanov [3] who extended the result of [2] to any number of variables and the recent paper of J.Hounie-P.Santiago [4].

In the present work, we prove local existence in dimension 2, of regular solutions for a class of quasi-linear differential operators of order 1 and complex coefficients. More precisely, we deal with equations of type

$$D_t u + c(t, x, u) D_x u = g(t, x, u)$$

where c and g are C^∞ of their arguments, analytic on u and with complex values.

We show this result under a "sign" hypothesis on the imaginary part of $c(t, x, u)$. This assumption will be precised later; it is a particular case of condition (P) of Nirenberg-Trêves [8] and is well adapted to the quasi-linear framework.

The proof is based on a classical iteration scheme and essentially uses the energy estimation (3-6) of [8].

1. Statement of the theorem - Remarks.

In the following, we will denote by Ω an open set of $\mathbb{R}^2$ containing the origin, and by (t, x) its current point; B_R will denote the ball of $\mathbb{C}$ with center at the origin and radius R.

We consider two functions $c(t, x, z)$ and $g(t, x, z)$ defined on $\Omega \times B_R$, with complex values, C^∞ of their arguments and analytic in z. We denote Imz the imaginary part of the complex number z and we suppose that it exists $r \in]0, R[$ such that the following hypothesis holds :

$$Im c(t,x,z) \geq 0 \quad for\ any \quad (t,x,z) \in \Omega \times \bar{B}_r \qquad (H)$$

We can now state our result.

Theorem : *In the previous context and under assumption (H), for any positive integer m, it exists a neighborhood V of the origin of $\mathbb{R}^2$ and a function $u \in C^m(V)$ satisfying, on V, the equation*

$$D_t u + c(t,x,u) D_x u = g(t,x,u).$$

Remarks : 1. The conclusion of the theorem is true near any point (t,x) of the neighborhood Ω of the origin. We have a local solvability result.

2. Condition (H) is a particular case of condition (P) of Nirenberg-Trêves (see for exemple [8]). It has the advantage to be stable, so well adapted to the non linear framework. Moreover, it guarantees the energy inequlity (3.6) of [8] (ie estimation (2.4) of the present paper) and insures, as remarked by N.Lerner [7] or Nirenberg-Trêves [8] that problems (2.3) and (2.5) are well posed. So it plays a central role in the iterative scheme used in the proof.

3. The dimension 2 framework adopted in this theorem, insures the regularity of the iteration solutions using traces regularity. In fact, in dimension larger than 2, these two traces does not suffice (under condition (H)) to control the regularity in all spectral directions.

4. This theorem generalises the result obtained by the author in [2]. It contains the hyperbolic quasi-linear case (taking $m = +\infty$). Moreover it is, in certain way, a "variant" to the local solvability theorem of Nirenberg-Trêves [8].

5. Let us note that an opertor of type $D_t + ia(t,x) D_x$ ($a \geq 0$ or $a \leq 0$) is generically the microlocal model of an operator verifying condition (P). This justifyes the structure choice of our operator.

2. Study of the linearised operator.

In what follows, for $T > 0$, $s \geq 0$ and $u \in L^\infty([-T,T], H^s)$, we will denote
$$\|u\|_{T,s} = \sup_{t \in [-T,T]} |u(t,.)|_s.$$

Let us first remark that, without loss of generality, we can replace the functions c and g of the statment, respectively by $\chi(t)\psi(x)g$ and $\psi(x)g$, where χ and ψ are suitable truncature functions.

More precisely, we take $\chi, \psi \in D(\mathbb{R})$, $\psi = 1$ for $|x| \leq 1$ and $\chi = 1$ for $|t| \leq \frac{T}{3}$, $Supp \chi \subset \{(t,x); \ |t| \leq \frac{T}{2}\}$.

We will continue to denote then, as in the theorem, by c and g, and we write $c = -b + ia$.

We take now a function u of class C^∞ in the band $[-T,T] \times \mathbb{R}$ and we examine the linearised operator on u.

$$L_u v = D_t v + c(t,x,u) D_x v = D_t v - b(t,x,u) D_x v + ia(t,x,u) D_x v. \tag{2.1}$$

We call $f(t,x)$ the unique C^∞ solution on $[-T,T] \times \mathbb{R}$ of the hyperbolic system
$$\begin{cases} D_t f - b(t,x,u) D_x f = 0 \\ f(0,x) = x \end{cases} \tag{2.2}$$

Clearly, for $|t|$ sufficiently small, the function $f(t,.)$ is bijective. Putting $\varphi(t,.) = (f(t,.))^{-1}$ and denoting by E_+ and E_- the two spectral projectors of the operator D_x, we can prove the following proposition.

Proposition 2.1 : *It exists $T > 0$ such that for any $g \in D(\mathbb{R}^2)$ and any v_+ and $v_- \in \bigcap_{s \geq 0} H^s(\mathbb{R})$, the system*

$$\begin{cases} L_u v = g & \text{in }]-T,T[\times \mathbb{R} \\ E_+(v \circ \varphi)(T) = v_+ \\ E_-(v \circ \varphi)(-T) = v_- \end{cases} \tag{2.3}$$

has a unique solution v in $\bigcap_{s \geq 0} C^0([-T,T], H^s)$ (so of class C^∞) verifying the energy estimate

$$||v||_{T,0} \leq c(u) \left(|v_+|_{L^2} + |v_-|_{L^2} + \int_{-T}^{T} |g(t,.)|_{L^2} dt \right) \tag{2.4}$$

Here, $c(u)$ is a positive constant depending only on $||u||_{T,s}$ for arbitrary $s > \frac{7}{2}$.

Proof : We choose T sufficiently small (depending on $||\nabla_x b(t,x,u)||_\infty$) to guarantee existence of the solution f off (2.2). We make then in system (2.3) the change of variables $(t', x') = (t, f(t,x))$.

It becomes
$$\begin{cases} D_{t'} \tilde{v} + i\tilde{a}(t', x', \tilde{u})(\frac{\partial f}{\partial x}) D_{x'} \tilde{v} = \tilde{g} \\ E_+(\tilde{v})(T) = v_+ \\ E_-(\tilde{v})(-T) = v_- \end{cases} \tag{2.5}$$

with $\tilde{v}(t', x') = v(t', f(t', x'))$, and the same for $\tilde{a}$ and $\tilde{g}$. Now, due to the energy inequality (3.6) of [8], for $T \leq T_0$, T_0 sufficiently small (depending on $||\nabla_x^2(\tilde{a}\frac{\partial f}{\partial x})||_\infty$), the system

(2.5) has a unique solution $\tilde{v} \in C^0([-T,T], L^2)$.

Morever, we have

$$\|v\|_{T,0} \le |v_+|_{L^2} + |v_-|_{L^2} + C(\tilde{a}\frac{\partial f}{\partial x}) \int_{-T}^{T} |\tilde{g}(t,.)|_{L^2} dt \qquad (2.6)$$

where $0 < C(\tilde{a}\frac{\partial f}{\partial x}) = C(\|\nabla^2_{x'}(\tilde{a}\frac{\partial f}{\partial x})\|_\infty) = C(\|u\|_{T,s})$, s being any real number choosen $> \frac{7}{2}$.

Here the L^2 norms are, of course, that of $L^2(\mathbb{R}_{x'})$.

Coming back to the (t,x) variables, the function $v(t,x) = \tilde{v}(t, f(t,x))$, is solution (for the moment in $C^0([0,T], L^2)$) of (2.3) and verifyes estimation (2.4).

On the other hand, for $-T < t' < -\frac{T}{2}$, $\tilde{v}$ verifyes $D_{t'}\tilde{v} = \tilde{g}$. So $E_-\tilde{v}(t') \in \bigcap_{s\ge 0} H^s$; this shows that $(t',x',0,\xi') \notin WF\tilde{v}$ for any $\xi' < 0$. Using Hörmander theorem for propagation of singularities ([6], proposition 3.5.1) we obtain that $(t',x',0,\xi') \notin WF\tilde{v}$ for any $t' \in]-T,T[$ and any $\xi' < 0$.

Finally, a similar argument for $\xi' > 0$ and the particular form of the function a (null for $|x|$ large) guarantee that $\tilde{v}$ so v are in $\bigcap_{s\ge 0} C^0([-T,T], H^s)$.

We can now give the proof of the theorem; it's the goal of next section.

3. Convergence of the iterative scheme.

We define this scheme by $u_0 = 0$ and for $n \ge 0$

$$\begin{cases} D_t u_{n+1} + c(t,x,u_n) D_x u_{n+1} = g(t,x,u_n) \\ E_+(u_{n+1} \circ \varphi_n)(T) = 0 \\ E_-(u_{n+1} \circ \varphi_n)(-T) = 0 \end{cases} \qquad (3.1)$$

$$\begin{cases} D_t f_n - b(t,x,u_n) D_x f_n = 0 \\ f_n(0,x) = x \end{cases} \qquad (3.2)$$

and $\varphi_n(t,.) = (f_n(t,.))^{-1}$.

Two essential facts have to be showed and insure the convergence of (3.1) : for fixed $s > \frac{7}{2}$, one can find $T > 0$ such that :

$$\text{The sequence } (u_n)_n \text{ is bounded in } L^\infty([-T,T], H^s) \qquad (3.3)$$

$$(u_n)_n \text{ is a Cauchy seqence in } L^\infty([-T,T], L^2) \qquad (3.4).$$

The rest of the work will be devoted to the proof of these two points.

We proceed by induction and suppose that $||u_j||_{T,s} \leq \alpha < r$ for $j = 0, 1, ..., n$.

Let us first remark that system (3.2) can be rewritten in the form

$$\begin{cases} \frac{\partial}{\partial t}(f_n - x) - b(t, x, u_n)\frac{\partial}{\partial x}(f_n - x) = b(t, x, u_n) \\ (f_n - x)(0, x) = 0 \end{cases} \quad (3.5)$$

We deduce for $k \leq s$

$$\begin{cases} \frac{\partial}{\partial t}(D_x^k(f_n - x)) - b(t, x, u_n)\frac{\partial}{\partial x}(D_x^k(f_n - x)) = D_x^k(b(u_n)) - [b(u_n), D_x^k]\frac{\partial}{\partial x}(f_n - x) \\ D_x^k(f_n - x)(0, x) = 0 \end{cases} \quad (3.6)$$

The hyperbolic energy inequality gives

$$|D_x^k(f_n - x)(t, .)|_{L^2} \leq C \int_{-T}^{T} \left| \left(D_x^k(b(u_n)) - [b(u_n), D_x^k]\frac{\partial}{\partial x}(f_n - x) \right)(t, .) \right|_{L^2} dt$$

where $C = C(||u_n||_{T,s'})$, with arbitrary $s' > \frac{3}{2}$. We can so consider that $C = C(\alpha)$. Applying then to the integral a classical estimate of "Cagliardo-Nirenberg" type for composed functions ([1], ch II), we obtain for $|t| \leq T$

$$|(f_n - x)(t, .)|_s \leq C \int_{-T}^{T} (1 + |u_n(t, .)|_s + |(f_n - x)(t, .)|_s) dt$$

ie
$$||f_n - x||_{T,s} \leq 2CT(1 + \alpha + ||f_n - x||_{T,s})$$

wich implies
$$||f_n - x||_{T,s} \leq C'T \quad (3.7)$$

if $T \leq T_0$, T_0 sufficiently small.

Denoting wihtout distinction by $\varphi_n(x)$ the functions $\varphi_n(T, x)$ and $\varphi_n(-T, x)$ (confusion has no consequence in the following), we deduce from (3.7) the existence of two constants C_1 and C_2 verifying

$$|\varphi_n' - 1| \leq C_1 T \quad \text{and for} \quad 2 \leq k \leq s \quad |\varphi_n^{(k)}| \leq C_2 T \quad (3.8)$$

Remark : According to the special from of the function $b = Rec$ (cutted by ψ), for $|x|$ large φ_n is identity function; so $\varphi_n' = 1$ and $\varphi_n^{(k)} = 0$ for $k \geq 2$.

We now come back to (3.1) and examine the system verifyed by $D_x^k u_{n+1}$, $0 \leq k \leq s$.

We have
$$D_t(D_x^k u_{n+1}) + c(t,x,u_n)D_x(D_x^k u_{n+1}) = D_x^k(g(u_n)) + [c(u_n), D_x^k]D_x u_{n+1} \quad (3.9)$$

On the other hand, expressing $D_x^k u_{n+1}\varphi_n$ in term of $D_x^k(u_{n+1}\varphi_n)$ and derivatives of order ≥ 2 of φ_n, and using estimates (3.8), we obtain

$$|E_\pm(D_x^k u_{n+1}\varphi_n)(\pm T)|_{L^2} \leq CT\|u_{n+1}\|_{T,s} \quad (3.10)$$

Combining this with energy inequality (2.4) and treating the second member of (3.9) like the one of (3.6), we can write for any $t \in [-T,T]$:

$$|D_x^k u_{n+1}(t,.)|_{L^2} \leq CT\|u_{n+1}\|_{T,s} + C'\int_{-T}^{T}(1+|u_{n+1}(t,.)|_s)dt \quad (3.11)$$

This yelds to
$$\|u_{n+1}\|_{T,s} \leq C''T(1+\|u_{n+1}\|_{T,s})$$

and so $\|u_{n+1}\|_{T,s} \leq \alpha$ if $T \leq T_1$, T_1 sufficiently small. At this step, point (3.3) is showed.

Let us now have a look on the system fulfilled by $u_{n+1} - u_n$. We precisely have

$$\begin{cases} D_t(u_{n+1}-u_n) + c(u_n)D_x(u_{n+1}-u_n) = g(u_n) - g(u_{n-1}) + \\ \qquad\qquad\qquad\qquad\qquad\qquad\qquad\qquad (c(u_{n-1})-c(u_n))D_x u_n \\ \qquad\qquad\qquad\qquad = G_n \\ E_\pm((u_{n+1}-u_n)\varphi_n)(\pm T) = E_\pm((u_n\varphi_{n-1})-(u_n\varphi_n))(\pm T) = h_{n\pm}. \end{cases} \quad (3.12)$$

Thanks to (3.3), we can easily estimate for $t \in [-T,T]$

$$|G_n(t,.)|_{L^2} \leq C\big(\sup_{|t|\leq T}|\nabla_x u_n(t,.)| + \sup_{|t|\leq T}|\nabla_x g(t,.)|\big)|(u_n-u_{n-1})(t,.)|_{L^2}$$

wich gives
$$|G_n(t,.)|_{L^2} \leq C'|(u_n-u_{n-1})(t,.)|_{L^2} \quad (3.13)$$

In the same way
$$|h_{n\pm}|_{L^2} \leq \sup_{|t|\leq T}|\nabla_x u_n(t,.)||\varphi_n - \varphi_{n-1}|_{L^2}$$
$$\leq C''|\varphi_n - \varphi_{n-1}|_{L^2}$$

But
$$|\varphi_n(x) - \varphi_{n-1}(x)|_{L^2} \leq C|\varphi_n(f_n(x)) - \varphi_{n-1}(f_n(x))|_{L^2}$$
$$= C|\varphi_{n-1}(f_{n-1}(x)) - \varphi_{n-1}(f_n(x))|_{L^2}$$
$$\leq C\sup|\nabla_x \varphi_{n-1}||f_n - f_{n-1}|_{L^2}$$

ie
$$|\varphi_n - \varphi_{n-1}|_{L^2} \leq C'''T |(u_n - u_{n-1})(\pm T,.)|_{L^2} \qquad (3.14)$$

the last estimation coming from a simple hyperbolic inequality applied to a system on $f_n - f_{n-1}$ and similar to (3.2). We combine then (3.13) and (3.14) with (2.4) and obtain

$$||u_{n+1} - u_n||_{T,0} \leq CT ||u_n - u_{n-1}||_{T,0} \qquad (3.15)$$

for $T \leq T_2$, T_2 sufficiently small. This ends the proof of point (3.4) and shows the theorem.

BIBLIOGRAPHY

[1] S. ALINAC et P. GERARD : Opérateurs pseudo-différentels et théorème de Nash-Moser. *Savoirs Actuels (1992)*.

[2] B. DEHMAN : Résolubilité locale pour des équations semi-linéaires complexes. *Canadian J. of Maths. Vol. XLII, N1, 1990, pp 126-140*.

[3] T. GRAMCHEV and P. POPIVANOV : Local solvability of semilinear partial differential equations. *Ann. Univ. Ferrara, 35, (1989), 147-154*.

[4] J. HOUNIE and P. SANTIAGO : On local solvability of semilinear equations. *Preprint*.

[5] L. HÖRMANDER : The analysis of linear partial differential operators. *Tome 2, Springer (1985)*.

[6] L. HÖRMANDER : On the existence and the regularity of linear pseudo-differential equations. *L'Enseignement Mathématique N18, Genève (1971)*.

[7] N. LERNER : Communication orale.

[8] L. NIRENBERG and F. TREVES : On local solvability of linear P.D.E. Part II. Sufficient conditions. *Comm. on Pure and Applied Maths. 23 (1970), 459-510*.

[9] P. SANTIAGO : Résolubilité locale pour des équations semi-linéaires complexes. *Thèse de Doctorat. Universidade Federal de Pernambuco. Recife. BRAZIL*.

[10] F. TREVES : Remarks on the integrability of first order complex P.D.E. *Journal of fuctional analysis. 106, 329-352, (1992)*.

B. DEHMAN
Département de mathématiques
Faculté des Sciences de Tunis
1060-TUNIS-TUNISIE

P DRÁBEK
Strongly nonlinear degenerated and singular elliptic problems[1]

1. Introduction

Let us consider the linear eigenvalue problem

$$-\Delta u - \lambda u = 0 \quad \text{in } \Omega,$$
$$u = 0 \quad \text{on } \partial\Omega, \tag{1.1}$$

on a bounded open set $\Omega \subset \mathbb{R}^n$. It is well known that problem (1.1) has the *least (first) eigenvalue* $\lambda_1 > 0$ and corresponding *eigenfunction* u_1 *positive* in Ω. This fact follows from a general result for abstract linear operators in semiordered spaces (see e.g. KREIN, RUTMAN [14], KRASNOSELSKIJ [13]). The situation is the same if we consider more general eigenvalue problem

$$-\text{div}\,(a(x)\nabla u) - \lambda b(x) u = 0 \quad \text{in } \Omega,$$
$$u = 0 \quad \text{on } \partial\Omega \tag{1.2}$$

with sufficiently smooth functions $a = a(x)$, $b = b(x)$ satisfying $0 < c_1 \le a(x) \le c_2$, $0 < c_3 \le b(x) \le c_4$ in Ω with some constants c_i, $i = 1, 2, 3, 4$. We may consider various generalizations of eigenvalue problems (1.1) and (1.2) and to prove an analogous results.

At first, we may consider eigenvalue problem (1.2) with *coefficients* $a(x), b(x)$ which *degenerate* (or have a *singularity*) in some sense. More precisely, if we assume that $b(x) \ge 0$ a.e. in Ω with strict inequality on the set of positive measure in Ω, $b(x) \in L^\infty(\Omega)$, $a(x) > 0$ a.e. in Ω, $a(x) \in L^1_{loc}(\Omega)$, $\frac{1}{a(x)} \in L^1_{loc}(\Omega)$, we can work in the weighted Sobolev space $W^{1,2}_0(a, \Omega)$ (with the weight $a(x)$) and to prove that the eigenvalue problem (1.2) has the least eigenvalue $\lambda_1 > 0$ with the corresponding nonnegative eigenfunction in Ω. The only difference from the nondegenerate case (1.1) consists in the introduction of the weighted Sobolev space and the concept of the weak solution of (1.2). *Secondly,* we may consider coefficients a and b in (1.2) dependent on the solution, i.e. the eigenvalue problem is of the form

---
[1] This work was supported by the Grant Agency of the Czech Republic under Grant No. 201/94/0008

$$-\text{div}\,(a(x,u(x))\nabla u) - \lambda b(x,u(x))u = 0 \quad \text{in } \Omega,$$
$$u = 0 \quad \text{on } \partial\Omega. \tag{1.3}$$

Note that the problem (1.3) is *nonlinear* and *nonpotential* in general. Let us mention the result of BOCCARDO [5] which for the eigenvalue problem (1.3) reads as follows: *Let $b(x,s) \equiv 1$ and $a(x,s)$ be a Carathéodory function satisfying $0 < c_1 \leq a(x,s) \leq c_2$ for any $s \in \mathbb{R}$ and for a.e. $x \in \Omega$. Then for each positive real number $R > 0$, we can find a positive eigenvalue $\lambda > 0$ with corresponding positive eigenfunction $u \in W_0^{1,2}(\Omega)$ such that $\|u\|_{L^2(\Omega)} = R$.*

In the third place, we may consider eigenvalue problems for *p-Laplacian:* $(p > 1)$:

$$-\text{div}\,(|\nabla u|^{p-2}\nabla u) - \lambda |u|^{p-2}u = 0 \quad \text{in } \Omega,$$
$$u = 0 \quad \text{on } \partial\Omega. \tag{1.4}$$

This problem should be treated as a *degenerated* (or *singular*) problem, where the degenerations (or singularities) are given by $|\nabla u|^{p-2}$ and $|u|^{p-2}$. The problem (1.4) is potential, the degenerations (or singularities) are, however, more complicated than in the case (1.2). There are several results concerning the existence of positive first eigenvalue and the corresponding positive eigenfunction of the problem (1.4). Let us mention e.g. papers of ANANE [2], BARLES [3], BHATTACHARYA [4], GARCÍA AZORERO, PERAL ALONSO [12], LINDQVIST [17], OTANI, TESHIMA [18] and others (see DRÁBEK [7] for other references).

The *aim of this work* is to study the eigenvalue problem which is a combination of preceding three cases. More precisely, ve study *nonhomogeneous degenerated quasilinear eigenvalue problem*

$$-\text{div}\,(a(x,u)|\nabla u|^{p-2}\nabla u) = \lambda b(x,u)|u|^{p-2}u \quad \text{in } \Omega,$$
$$u = 0 \quad \text{on } \partial\Omega. \tag{1.5}$$

At first we study homogeneous degenerated problem associated with (1.5). Secondly, using the results for homogeneous problem, apriori estimate in $L^\infty(\Omega)$ and the Schauder fixed point theorem we prove the assertion for (1.5) which is analogous to that proved by BOCCARDO [5].

This result is applied in order to study the *solvability of quasilinear elliptic boundary value problem* (BVP)

$$-\text{div}(a(x,u)|\nabla u|^{p-2}\nabla u) = \lambda b(x,u)|u|^{p-2}u + f(x,u) \quad \text{in } \Omega,$$
$$u = 0 \quad \text{on } \partial\Omega, \tag{1.6}$$

where $f(x,s)$ is a nonlinear function satisfying appropriate growth conditions. The main tool is a functional-theoretical approach based on the degree theory for a generalized monotone mappings (see BROWDER, PETRYSHIN [6] and SKRYPNIK [19]). We prove also a variant of the *weak maximum principle* and generalize thus the results of FLECKINGER, HERNANDEZ, de THÉLIN [10].

2. Basic function spaces

Let us suppose that Ω is an open bounded subset of n-dimensional Euclidean space $\mathbb{R}^n$, $p > 1$ is an arbitrary real number and w is a *weight function* (i.e. positive and measurable) in Ω. Assume that

$$w \in L^1_{loc}(\Omega) \text{ and } \frac{1}{w} \in L^{\frac{1}{p-1}}_{loc}(\Omega). \tag{2.1}$$

Let us define the *weigted Sobolev space* $W^{1,p}(w,\Omega)$ as the set of all real valued functions u defined in Ω for which

$$\|u\|_{1,p,w} = \left(\int_\Omega |u|^p dx + \int_\Omega w |\nabla u|^p dx \right)^{\frac{1}{p}} < \infty. \tag{2.2}$$

It follows from (2.1) that $W^{1,p}(w,\Omega)$ is a *reflexive Banach space* and that $W^{1,p}_0(w,\Omega)$ is well defined as a closure of $C^\infty_0(\Omega)$ in $W^{1,p}(w,\Omega)$ with respect to the norm $\|\cdot\|_{1,p,w}$ (see e.g. KUFNER, SÄNDIG[16]).

Let $s \geq \frac{1}{p-1}$ be a real number. A simple application of the Hölder inequality yields that the *continuous imbedding*

$$W^{1,p}(w,\Omega) \hookrightarrow W^{1,p_1}(\Omega) \tag{2.3}$$

holds provided

$$\frac{1}{w} \in L^s(\Omega) \text{ and } p_1 = \frac{ps}{s+1}.$$

It follows from (2.3) and from the Sobolev imbedding theorem (see e.g. ADAMS [1], KUFNER, JOHN, FUČÍK [15]) that for $s+1 \leq ps < n(s+1)$ we have

$$W^{1,p}_0(w,\Omega) \hookrightarrow W^{1,p_1}_0(\Omega) \hookrightarrow L^q(\Omega), \tag{2.4}$$

where $1 \leq q = \frac{np_1}{n-p_1} = \frac{nps}{n(s+1)-ps}$, and for $ps \geq n(s+1)$ the imbedding (2.4) holds with arbitrary $1 \leq q < \infty$.

Moreover, the *compact imbedding*

$$W^{1,p}_0(w,\Omega) \hookrightarrow\hookrightarrow L^r(\Omega)$$

holds provided $1 \leq r < q$.

An easy calculation yields that $s > \frac{n}{p}$ implies $q > p$. In particular, we have

$$W^{1,p}_0(w,\Omega) \hookrightarrow\hookrightarrow L^{p+\eta}(\Omega) \tag{2.5}$$

for $0 \leq \eta < q - p$, provided

$$\frac{1}{w} \in L^s(\Omega) \text{ and } s \in \left(\frac{n}{p}, +\infty\right) \cap \left[\frac{1}{p-1}, +\infty\right). \tag{2.6}$$

We prove a variant of Friedrichs inequality in our weighted Sobolev space.

In what follows we will always assume that (2.6) is fulfilled. Let $u \in C_0^\infty(\Omega)$. Then due to $q > p$ and the imbedding $W_0^{1,p_1}(\Omega) \hookrightarrow L^q(\Omega)$ we have

$$\left(\int_\Omega |u|^p dx\right)^{\frac{1}{p}} \leq c_1 \left(\int_\Omega |u|^q dx\right)^{\frac{1}{q}} \leq c_2 \left(\int_\Omega [|u|^{p_1} + |\nabla u|^{p_1}] dx\right)^{\frac{1}{p_1}}. \tag{2.7}$$

The Friedrichs inequality in $W_0^{1,p_1}(\Omega)$ yields

$$\left(\int_\Omega [|u|^{p_1} + |\nabla u|^{p_1}] dx\right)^{\frac{1}{p_1}} \leq c_3 \left(\int_\Omega |\nabla u|^{p_1} dx\right)^{\frac{1}{p_1}}. \tag{2.8}$$

Using the Hölder inequality we obtain

$$\left(\int_\Omega |\nabla u|^{p_1} dx\right)^{\frac{1}{p_1}} = \left(\int_\Omega |\nabla u|^{p_1} w^{\frac{p_1}{p}} \frac{1}{w^{\frac{p_1}{p}}} dx\right)^{\frac{1}{p_1}} \leq$$

$$\leq \left(\int_\Omega w |\nabla u|^p dx\right)^{\frac{1}{p}} \left(\int_\Omega \frac{1}{w^{\frac{p_1}{p} \cdot \frac{p}{p-p_1}}} dx\right)^{\frac{p-p_1}{p} \cdot \frac{1}{p_1}} \leq \tag{2.9}$$

$$\leq \left(\int_\Omega \frac{1}{w^s} dx\right)^{\frac{1}{ps}} \left(\int_\Omega w|\nabla u|^p dx\right)^{\frac{1}{p}}$$

(see (2.4) for the relation between s, p and p_1). It follows from (2.7) – (2.9) that

$$\int_\Omega |u|^p dx \leq c_4 \int_\Omega w|\nabla u|^p dx$$

with a constant $c_4 > 0$ independent of $u \in C_0^\infty(\Omega)$. Hence the norm

$$\|u\|_w = \left(\int_\Omega w|\nabla u|^p dx\right)^{\frac{1}{p}}$$

on the space $W_0^{1,p}(w, \Omega)$ is equivalent to the norm $\|\cdot\|_{1,p,w}$ defined by (2.2).

Let us assume that $\tilde{w}$ is a weight function defined in Ω and satisfying inequalities

$$c_5 w(x) \leq \tilde{w}(x) \leq c_6 w(x) \tag{2.10}$$

for a.e. $x \in \Omega$ with some constants $c_6 \geq c_5 > 0$. Then obviously

$$W_0^{1,p}(\tilde{w}, \Omega) = W_0^{1,p}(w, \Omega)$$

and the norms $\|\cdot\|_{\tilde{w}}$ and $\|\cdot\|_w$ are *equivalent* on $W_0^{1,p}(w, \Omega)$. It follows from Clarkson's inequality (see KUFNER, JOHN, FUČÍK [15]) that $W_0^{1,p}(w, \Omega)$ is *uniformly convex Banach space* with respect to the norm $\|\cdot\|_{\tilde{w}}$ for any $\tilde{w}$ satisfying (2.10).

Lemma 2.1 *Let $p \geq 2$. Then*

$$|t_2|^p - |t_1|^p \geq p|t_1|^{p-2}t_1(t_2 - t_1) + \frac{|t_2 - t_1|^p}{2^{p-1} - 1} \qquad (2.11)$$

for all points t_1 and t_2 in $\mathbb{R}^n$.
Let $1 < p < 2$. Then

$$|t_2|^p - |t_1|^p \geq p|t_1|^{p-2}t_1(t_2 - t_1) + \frac{3p(p-1)}{16} \frac{|t_2 - t_1|^2}{(|t_1| + |t_2|)^{2-p}} \qquad (2.12)$$

for all points t_1 and t_2 in $\mathbb{R}^n$.

Proof of this lemma is based on Clarkson's inequality and can be found in LINDQVIST [17].

Remark 2.1 It follows from (2.11) and (2.12) that the inequality

$$|t_2|^p - |t_1|^p > p|t_1|^{p-2}t_1(t_2 - t_1) \qquad (2.13)$$

holds for any $t_1, t_2 \in \mathbb{R}^n$, $t_1 \neq t_2$ and for any $p > 1$. Note that inequality (2.13) is just a restating of the strict convexity of the mapping $t \mapsto |t|^p$ and can be proved independently of (2.11) and (2.12).

3. Homogeneous eigenvalue problem

Let us suppose that w is the weight function satisfying (2.1) and (2.6). Let $a(x), b(x)$ be measurable functions satisfying

$$\frac{w(x)}{c_1} \leq a(x) \leq c_1 w(x), \qquad (3.1)$$

$$0 \leq b(x) \qquad (3.2)$$

for a.e. $x \in \Omega$ with some constant $c_1 > 1$, and $b(x) \in L^{\frac{q^*}{q^*-p}}(\Omega)$ for $p < q^* < q$, $b(x) \in L^\infty(\Omega)$ for $q^* = p$ (see (2.4) for q). Moreover, let

$$\text{meas}\{x \in \Omega; b(x) > 0\} > 0. \qquad (3.3)$$

Further we will assume that $p < q^* < q$. The proofs can be performed in the same way also in the case $q^* = p$.

Let us consider *homogeneous eigenvalue problem*

$$\begin{aligned} -\operatorname{div}(a(x)|\nabla u|^{p-2}\nabla u) &= \lambda b(x)|u|^{p-2}u \text{ in } \Omega, \\ u &= 0 \text{ on } \partial\Omega. \end{aligned} \qquad (3.4)$$

Definition 3.1 We will say that $\lambda \in \mathbb{R}$ is the *eigenvalue* and $u \in W_0^{1,p}(w,\Omega)$, $u \neq 0$, is the corresponding *eigenfunction* of the eigenvalue problem (3.4) if

$$\int_\Omega a(x)|\nabla u|^{p-2}\nabla u \nabla \varphi\, dx = \lambda \int_\Omega b(x)|u|^{p-2}u\varphi\, dx \tag{3.5}$$

holds for any $\varphi \in W_0^{1,p}(w,\Omega)$.

Lemma 3.1 *There exists the least (the first) eigenvalue $\lambda_1 > 0$ and at least one corresponding eigenfunction $u_1 \geq 0$ a.e. in Ω ($u_1 \not\equiv 0$) of the eigenvalue problem (3.4).*

Proof Set

$$\lambda_1 = \inf\{\int_\Omega a(x)|\nabla v|^p dx;\ \int_\Omega b(x)|v|^p dx = 1\}.$$

Obviously $\lambda_1 \geq 0$. Let (v_n) be the minimizing sequence for λ_1, i.e.

$$\int_\Omega b(x)|v_n|^p dx = 1 \text{ and } \int_\Omega a(x)|\nabla v_n|^p dx = \lambda_1 + \delta_n, \tag{3.6}$$

with $\delta_n \to 0_+$ for $n \to \infty$. It follows from (3.6) that $\|v_n\|_a \leq c_2$, with $c_2 > 0$ independent of n. The reflexivity of $W_0^{1,p}(w,\Omega)$ yields the weak convergence $v_n \rightharpoonup u_1$ in $W_0^{1,p}(w,\Omega)$ for some u_1 (at least for some subsequence of (v_n)). The compact imbedding $W_0^{1,p}(w,\Omega) \hookrightarrow\hookrightarrow L^{q^*}(\Omega)$ implies the strong convergence $v_n \to u_1$ in $L^{q^*}(\Omega)$. It follows from (3.2),(3.6), from the Minkowski and the Hölder inequality

$$1 = \lim_{n\to\infty}\left(\int_\Omega b(x)|v_n|^p dx\right)^{\frac{1}{p}} \leq \lim_{n\to\infty}\left(\int_\Omega b(x)|v_n - u_1|^p dx\right)^{\frac{1}{p}} + \left(\int_\Omega b(x)|u_1|^p dx\right)^{\frac{1}{p}} \leq$$

$$\leq \lim_{n\to\infty}\left(\int_\Omega (b(x))^{\frac{q^*}{q^*-p}}dx\right)^{\frac{q^*-p}{pq^*}} \cdot \left(\int_\Omega |v_n - u_1|^{q^*} dx\right)^{\frac{1}{q^*}} + \left(\int_\Omega b(x)|u_1|^p dx\right)^{\frac{1}{p}} =$$

$$= \left(\int_\Omega b(x)|u_1|^p dx\right)^{\frac{1}{p}},$$

and analogously

$$\left(\int_\Omega b(x)|u_1|^p dx\right)^{\frac{1}{p}} \leq$$

$$\leq \lim_{n\to\infty}\left(\int_\Omega (b(x))^{\frac{q^*}{q^*-p}}dx\right)^{\frac{q^*-p}{pq^*}} \cdot \left(\int_\Omega |u_1 - v_n|^{q^*} dx\right)^{\frac{1}{q^*}} + \lim_{n\to\infty}\left(\int_\Omega b(x)|v_n|^p dx\right)^{\frac{1}{p}} = 1.$$

Hence

$$\int_\Omega b(x)|u_1|^p dx = 1.$$

In particular, $u_1 \not\equiv 0$. The property of weakly convergent sequence (v_n) in $W_0^{1,p}(w,\Omega)$ yields

$$\lambda_1 \leq \int_\Omega a(x)|\nabla u_1|^p dx = \|u_1\|_a^p \leq \liminf_{n\to\infty} \|v_n\|_a^p =$$

$$= \liminf_{n\to\infty} \int_\Omega a(x)|\nabla v_n|^p dx = \liminf_{n\to\infty}(\lambda_1 + \delta_n) = \lambda_1,$$

i.e.

$$\lambda_1 = \int_\Omega a(x)|\nabla u_1|^p dx. \tag{3.7}$$

It follows from (3.7) that $\lambda_1 > 0$ and it is easy to see that λ_1 is the least eigenvalue of (3.4) with the corresponding eigenfunction u_1.

Moreover, if u is some eigenfunction corresponding to λ_1 then $|u|$ is also the eigenfunction corresponding to λ_1. Hence we can suppose that $u_1 \geq 0$ a.e. in Ω. ∎

Remark 3.1 It follows from the proof of Lemma 3.1 that $v_n \rightharpoonup u_1$ in $W_0^{1,p}(w, \Omega)$ and $\|v_n\|_a \to \|u_1\|_a$. The uniform convexity of $W_0^{1,p}(w, \Omega)$ (see Section 2) then implies the *strong convergence* $v_n \to u_1$ in $W_0^{1,p}(w, \Omega)$.

Lemma 3.2 *Let $u \in W_0^{1,p}(w, \Omega), u \geq 0$ a.e. in Ω, be the eigenfunction corresponding to the first eigenvalue $\lambda_1 > 0$ of the eigenvalue problem (3.4). Then $u \in L^\infty(\Omega)$.*

Proof For $M > 0$ define

$$v_M(x) = \inf\{u(x), M\} \in W_0^{1,p}(w, \Omega) \cap L^\infty(\Omega).$$

Let us choose $\varphi = v_M^{\kappa p+1} (\kappa \geq 0)$ in

$$\int_\Omega a(x)|\nabla u|^{p-2} \nabla u \nabla \varphi dx = \lambda_1 \int_\Omega b(x)|u|^{p-2} u\varphi dx. \tag{3.8}$$

Obviously $\varphi \in W_0^{1,p}(w, \Omega) \cap L^\infty(\Omega)$. It follows from (3.8) that

$$(\kappa p + 1) \int_\Omega a(x) v_M^{\kappa p} |\nabla v_M|^p dx = \lambda_1 \int_\Omega b(x) u^{p-1} v_M^{\kappa p+1} dx. \tag{3.9}$$

Due to (3.1) and the imbedding $W_0^{1,p}(w, \Omega) \hookrightarrow L^q(\Omega)$ we have

$$(\kappa p + 1) \int_\Omega a(x) v_M^{\kappa p} |\nabla v_M|^p dx \geq$$
$$\geq \frac{\kappa p + 1}{c_1} \int_\Omega w(x) v_M^{\kappa p} |\nabla v_M|^p dx = \tag{3.10}$$
$$= \frac{\kappa p + 1}{c_1(\kappa+1)^p} \int_\Omega w(x) |\nabla(v_M^{\kappa+1})|^p dx \geq c_3 \frac{(\kappa p + 1)}{(\kappa+1)^p} \left(\int_\Omega (v_M^{\kappa+1})^q dx\right)^{\frac{p}{q}}.$$

Hence it follows from (3.2), (3.8), (3.9), (3.10) and the Hölder inequality that

$$\left(\int_\Omega v_M^{(\kappa+1)q} dx\right)^{\frac{p}{q}} \leq \frac{1}{c_3}\frac{(\kappa+1)^p}{\kappa p+1}\int_\Omega b(x)u^{p-1}v_M^{\kappa p+1} dx \leq$$

$$\leq \frac{1}{c_3}\frac{(\kappa+1)^p}{\kappa p+1}\left(\int_\Omega b(x)^{\frac{q^*}{q^*-p}} dx\right)^{\frac{q^*-p}{q^*}}\cdot\left(\int_\Omega u^{(\kappa+1)q^*} dx\right)^{\frac{p}{q^*}}. \quad (3.11)$$

Due to the assumptions on $b(x)$ we obtain formally

$$\left(\int_\Omega v_M^{(\kappa+1)q} dx\right)^{\frac{p}{q}} \leq c_4 \frac{(\kappa+1)^p}{\kappa p+1}\left(\int_\Omega u^{(\kappa+1)q^*} dx\right)^{\frac{p}{q^*}}, \quad (3.12)$$

i.e.

$$\|v_M\|_{(\kappa+1)q} \leq c_5^{\frac{1}{\kappa+1}}\left[\frac{\kappa+1}{(\kappa p+1)^{\frac{1}{p}}}\right]^{\frac{1}{\kappa+1}}\|u\|_{(\kappa+1)q^*} \quad (3.13)$$

with $c_5 = c_4^{\frac{1}{p}}$. Since $u \in L^q(\Omega)$, we can choose κ_1 in (3.13) such that $(\kappa_1+1)q^* = q$, i.e. $\kappa_1 = \frac{q}{q^*} - 1$. Then we have

$$\|v_M\|_{(\kappa_1+1)q} \leq c_5^{\frac{1}{\kappa_1+1}}\left[\frac{\kappa_1+1}{(\kappa_1 p+1)^{\frac{1}{p}}}\right]^{\frac{1}{\kappa_1+1}}\|u\|_q \quad (3.14)$$

for any $M > 0$. Due to $u(x) = \lim_{M\to\infty} v_M(x)$, the Fatou lemma and (3.14) imply

$$\|u\|_{(\kappa_1+1)q} \leq c_5^{\frac{1}{\kappa_1+1}}\left[\frac{\kappa_1+1}{(\kappa_1 p+1)^{\frac{1}{p}}}\right]^{\frac{1}{\kappa_1+1}}\|u\|_q. \quad (3.15)$$

Hence, we can choose κ_2 in (3.13) such that $(\kappa_2+1)q^* = (\kappa_1+1)q = \frac{(q)^2}{q^*}$ and repeating the same argument we get

$$\|u\|_{(\kappa_2+1)q} \leq c_5^{\frac{1}{\kappa_2+1}}\left[\frac{\kappa_2+1}{(\kappa_2 p+1)^{\frac{1}{p}}}\right]^{\frac{1}{\kappa_2+1}}\|u\|_{(\kappa_1+1)q}.$$

By induction we obtain

$$\|u\|_{(\kappa_n+1)q} \leq c_5^{\frac{1}{\kappa_n+1}}\left[\frac{\kappa_n+1}{(\kappa_n p+1)^{\frac{1}{p}}}\right]^{\frac{1}{\kappa_n+1}}\|u\|_{(\kappa_{n-1}+1)q} \quad (3.16)$$

for any $n \in \mathbb{N}$, where $(\kappa_n+1) = \left(\frac{q}{q^*}\right)^n$. It follows from (3.15), (3.16) that

$$\|u\|_{(\kappa_n+1)q} \leq c_5^{\sum_{k=1}^n \frac{1}{\kappa_k+1}}\left[\left(\frac{\kappa_1+1}{(\kappa_1 p+1)^{\frac{1}{p}}}\right)^{\frac{1}{\sqrt{\kappa_1+1}}}\right]^{\frac{1}{\sqrt{\kappa_1+1}}}$$

119

$$\left[\left(\frac{\kappa_2+1}{(\kappa_2 p+1)^{\frac{1}{p}}}\right)^{\frac{1}{\sqrt{\kappa_2+1}}}\right]^{\frac{1}{\sqrt{\kappa_2+1}}} \cdots \left[\left(\frac{\kappa_n+1}{(\kappa_n p+1)^{\frac{1}{p}}}\right)^{\frac{1}{\sqrt{\kappa_n+1}}}\right]^{\frac{1}{\sqrt{\kappa_n+1}}} \|u\|_q.$$

Since $\left[\frac{y+1}{(yp+1)^{\frac{1}{p}}}\right]^{\frac{1}{\sqrt{y+1}}} > 1$ and $\lim_{y\to\infty}\left[\frac{y+1}{(yp+1)^{\frac{1}{p}}}\right]^{\frac{1}{\sqrt{y+1}}} = 1$, there exists $c_6 > 1$ (independent of κ_n) such that

$$\|u\|_{(\kappa_n+1)q} \leq c_5^{\sum_{k=1}^{n}\frac{1}{\kappa_k+1}} c_6^{\sum_{k=1}^{n}\frac{1}{\sqrt{\kappa_k+1}}} \|u\|_q. \tag{3.17}$$

But $\sum_{k=1}^{n}\frac{1}{\kappa_k+1} = \sum_{k=1}^{n}\left(\frac{q^*}{q}\right)^k$, $\sum_{k=1}^{n}\frac{1}{\sqrt{\kappa_k+1}} = \sum_{k=1}^{n}\left(\sqrt{\frac{q^*}{q}}\right)^k$, and $\frac{q^*}{p} < \sqrt{\frac{q^*}{p}} < 1$. Hence there exists a constant $c_7 > 0$ (independent of $n \in \mathbb{N}$) such that letting $n \to \infty$ in (3.17) we get

$$\|u\|_\infty \leq c_7 \|u\|_q \tag{3.18}$$

(see e.g. ADAMS [1]). It follows from (3.17) and (3.18) that $u \in L^r(\Omega)$ for all $1 \leq r \leq \infty$.

∎

Proposition 3.1 *There exists precisely one nonnegative eigenfunction u_1, $\|u_1\|_{q^*} = 1$, corresponding to the first eigenvalue $\lambda_1 > 0$ of the eigenvalue problem (3.4).*

Proof Due to the variational characterization of λ_1 the function $u \in W_0^{1,p}(w,\Omega)$ is an eigenfunction corresponding to λ_1 if and only if

$$\int_\Omega a(x)|\nabla u|^p dx - \lambda_1 \int_\Omega b(x)|u|^p dx = 0 =$$

$$= \inf_{v \in W_0^{1,p}(w,\Omega)} \left[\int_\Omega a(x)|\nabla v|^p dx - \lambda_1 \int_\Omega b(x)|v|^p dx\right].$$

It follows from here that if $u_1, u_2 \in W_0^{1,p}(w,\Omega)$ are two eigenfunctions corresponding to λ_1 then also

$$v_1(x) = \max_{x\in\Omega}\{u_1(x), u_2(x)\}, \quad v_2(x) = \min_{x\in\Omega}\{u_1(x), u_2(x)\}$$

are eigenfunctions corresponding to λ_1 provided that $v_2 \not\equiv 0$. Really, we have $v_1, v_2 \in W_0^{1,p}(w,\Omega)$ and

$$\int_\Omega a(x)|\nabla v_1|^p dx - \lambda_1 \int_\Omega b(x)|v_1|^p dx + \int_\Omega a(x)|\nabla v_2|^p dx - \lambda_1 \int_\Omega b(x)|v_2|^p dx =$$

$$= \int_\Omega a(x)|\nabla u_1|^p dx - \lambda_1 \int_\Omega b(x)|u_1|^p dx + \int_\Omega a(x)|\nabla u_2|^p dx - \lambda_1 \int_\Omega b(x)|u_2|^p dx.$$

Hence

$$\int_\Omega a(x)|\nabla v_1|^P dx - \lambda_1 \int_\Omega b(x)|v_1|^P dx = \int_\Omega a(x)|\nabla v_2|^P dx - \lambda_1 \int_\Omega b(x)|v_2|^P dx = 0.$$

Let $u_1 \geq 0$ and $u_2 \geq 0$ be two eigenfunctions corresponding to λ_1 such that $u_1 \not\equiv u_2$, $\min_{x \in \Omega}\{u_1(x), u_2(x)\} \not\equiv 0$ and

$$\|u_1\|_{q^*} = \|u_2\|_{q^*} = 1.$$

Denote by $v_3(x) = k_1 v_2(x) = k_1 \min_{x \in \Omega}\{u_1(x), u_2(x)\}$, where $k_1 > 0$ is chosen in such a way that

$$\|v_3\|_{q^*} = 1.$$

Then $v_3 \in W_0^{1,p}(w, \Omega)$ is again the eigenfunction corresponding to λ_1 such that $v_3 \not\equiv u_1$. Moreover

$$\{x \in \Omega; u_1(x) = 0\} \subseteq \{x \in \Omega; v_3(x) = 0\}.$$

Set $v_5(x) = k_2 v_4(x) = k_2 \max_{x \in \Omega}\{u_1(x), v_3(x)\}$, where $k_2 > 0$ is chosen such that

$$\|v_5\|_{q^*} = 1.$$

Then $v_5 \in W_0^{1,p}(w, \Omega)$ is the eigenfunction corresponding to λ_1 such that $v_5 \not\equiv u_1$ and

$$\{x \in \Omega; v_5(x) = 0\} = \{x \in \Omega; u_1(x) = 0\}.$$

Let, now, $u_1 \geq 0$ and $u_2 \geq 0$ be two eigenfunctions corresponding to λ_1 such that $u_1 \not\equiv u_2$, $\|u_1\|_{q^*} = \|u_2\|_{q^*} = 1$ and

$$\min_{x \in \Omega}\{u_1(x), u_2(x)\} \equiv 0.$$

Denote $\tilde{u}_1 = k_3 \max\{u_1(x), u_2(x)\}$, where $0 < k_3 < 1$ is chosen such that

$$\|\tilde{u}_1\|_{q^*} = 1,$$

and $\tilde{u}_2 = k_4 \max\{u_1(x), \tilde{u}_1(x)\}$, where $0 < k_4 < 1$ is such that

$$\|\tilde{u}_2\|_{q^*} = 1.$$

Then $\tilde{u}_1$ and $\tilde{u}_2$ are the eigenfunctions corresponding to λ_1 such that $\tilde{u}_1 \not\equiv \tilde{u}_2$ and

$$\{x \in \Omega; \tilde{u}_1 = 0\} = \{x \in \Omega; \tilde{u}_2 = 0\}.$$

We will prove the assertion of proposition via contradiction. Due to the argument presented above we assume that $u \geq 0$ and $v \geq 0$ are eigenfunctions corresponding to λ_1 such that

$$\|u\|_{q^*} = \|v\|_{q^*} = 1, \quad u \not\equiv v, \tag{3.19}$$

and which vanish in Ω on the same set (almost everywhere in the sense of the Lebesgue measure). Then

$$\int_\Omega a(x)|\nabla u|^{p-2}\nabla u \nabla \varphi dx = \lambda_1 \int_\Omega b(x)|u|^{p-2}u\varphi dx \tag{3.20}$$

for any $\varphi \in W_0^{1,p}(w,\Omega)$, and

$$\int_\Omega a(x)|\nabla v|^{p-2}\nabla v \nabla \psi dx = \lambda_1 \int_\Omega b(x)|v|^{p-2}v\psi dx \tag{3.21}$$

for any $\psi \in W_0^{1,p}(w,\Omega)$. For $\varepsilon > 0$ set

$$u_\varepsilon = u + \varepsilon \text{ and } v_\varepsilon = v + \varepsilon.$$

Substitute

$$\varphi = \frac{u_\varepsilon^p - v_\varepsilon^p}{u_\varepsilon^{p-1}}$$

into (3.20) and

$$\psi = \frac{v_\varepsilon^p - u_\varepsilon^p}{v_\varepsilon^{p-1}}$$

into (3.21). Since $\frac{u_\varepsilon}{v_\varepsilon}, \frac{v_\varepsilon}{u_\varepsilon} \in L^\infty(\Omega)$ and

$$\nabla \varphi = [1 + (p-1)(\frac{v_\varepsilon}{u_\varepsilon})^p]\nabla u - p(\frac{v_\varepsilon}{u_\varepsilon})^{p-1}\nabla v,$$

$$\nabla \psi = [1 + (p-1)(\frac{u_\varepsilon}{v_\varepsilon})^p]\nabla v - p(\frac{u_\varepsilon}{v_\varepsilon})^{p-1}\nabla u,$$

we have $\varphi, \psi \in W_0^{1,p}(w,\Omega)$. Adding (3.20) and (3.21) (with φ and ψ chosen above) we obtain

$$\int_\Omega a(x)\{[1+(p-1)(\frac{v_\varepsilon}{u_\varepsilon})^p]|\nabla u|^p + [1+(p-1)(\frac{u_\varepsilon}{v_\varepsilon})^p]|\nabla v|^p\}dx -$$

$$- \int_\Omega a(x)\{p(\frac{v_\varepsilon}{u_\varepsilon})^{p-1}|\nabla u|^{p-2}\nabla u \nabla v + p(\frac{u_\varepsilon}{v_\varepsilon})^{p-1}|\nabla v|^{p-2}\nabla v \nabla u\}dx =$$

$$= \lambda_1 \int_\Omega b(x)[(\frac{u}{u_\varepsilon})^{p-1} - (\frac{v}{v_\varepsilon})^{p-1}](u_\varepsilon^p - v_\varepsilon^p)dx.$$

Since $|\nabla \log u_\varepsilon| = \frac{|\nabla u|}{u_\varepsilon}$, the last equality is equivalent to

122

$$\int_\Omega a(x)(u_\varepsilon^p - v_\varepsilon^p)[|\nabla \log u_\varepsilon|^p - |\nabla \log v_\varepsilon|^p] dx -$$

$$- \int_\Omega a(x) p v_\varepsilon^p |\nabla \log u_\varepsilon|^{p-2} \nabla \log u_\varepsilon (\nabla \log v_\varepsilon - \nabla \log u_\varepsilon) dx -$$

$$- \int_\Omega a(x) p u_\varepsilon^p |\nabla \log v_\varepsilon|^{p-2} \nabla \log v_\varepsilon (\nabla \log u_\varepsilon - \nabla \log v_\varepsilon) dx = \quad (3.22)$$

$$= \lambda_1 \int_\Omega b(x)[(\frac{u}{u_\varepsilon})^{p-1} - (\frac{v}{v_\varepsilon})^{p-1}](u_\varepsilon^p - v_\varepsilon^p) dx.$$

Let $p \geq 2$. We use (2.11) in order to estimate the left hand side of (3.22) (we set at first $t_1 = \nabla \log u_\varepsilon, t_2 = \nabla \log v_\varepsilon$ and then $t_1 = \nabla \log v_\varepsilon, t_2 = \nabla \log u_\varepsilon$). We obtain

$$\lambda_1 \int_\Omega b(x)[(\frac{u}{u_\varepsilon})^{p-1} - (\frac{v}{v_\varepsilon})^{p-1}](u_\varepsilon^p - v_\varepsilon^p) dx \geq$$

$$\geq \frac{1}{2^{p-1} - 1} \int_\Omega a(x) |\nabla \log u_\varepsilon - \nabla \log v_\varepsilon|^p (u_\varepsilon^p + v_\varepsilon^p) dx = \quad (3.23)$$

$$= \frac{1}{2^{p-1} - 1} \int_\Omega a(x)(\frac{1}{v_\varepsilon^p} + \frac{1}{u_\varepsilon^p}) |v_\varepsilon \nabla u - u_\varepsilon \nabla v|^p dx \geq 0.$$

Let $1 < p < 2$. We use (2.12) in order to estimate the left hand side of (3.22) (similarly as above) and we obtain

$$\lambda_1 \int_\Omega b(x)[(\frac{u}{u_\varepsilon})^{p-1} - (\frac{v}{v_\varepsilon})^{p-1}](u_\varepsilon^p - v_\varepsilon^p) dx \geq$$

$$\geq \frac{3p(p-1)}{16} \int_\Omega a(x)(\frac{1}{u_\varepsilon^p} + \frac{1}{v_\varepsilon^p}) \frac{|v_\varepsilon \nabla u - u_\varepsilon \nabla v|^2}{(v_\varepsilon|\nabla u| + u_\varepsilon|\nabla v|)^{2-p}} dx \geq 0. \quad (3.24)$$

We have $u, v \in L^\infty(\Omega)$ (see Lemma 3.2) and

$$\frac{u}{u_\varepsilon} \to 1, \quad \frac{v}{v_\varepsilon} \to 1 \quad (\varepsilon \to 0_+) \quad (3.25)$$

a.e. in Ω where $u > 0$ and $v > 0$, respectively;

$$\frac{u}{u_\varepsilon} = 0, \quad \frac{v}{v_\varepsilon} = 0 \text{ (for any } \varepsilon > 0) \quad (3.26)$$

elsewhere (since u and v vanish on the same set in Ω). Hence it follows from (3.25), (3.26) and the Lebesgue theorem that for any $p, 1 < p < \infty$

$$\lambda_1 \int_\Omega b(x)[(\frac{u}{u_\varepsilon})^{p-1} - (\frac{v}{v_\varepsilon})^{p-1}](u_\varepsilon^p - v_\varepsilon^p) dx \to 0 \quad (\varepsilon \to 0_+).$$

It follows from here, (3.23), (3.24) and from the Fatou lemma that

$$|v \nabla u - u \nabla v| = 0 \text{ a.e. in } \Omega$$

for any $1 < p < \infty$. Hence there exists a constant $k > 0$ such that $u = kv$ a.e. in Ω. But (3.19) yields $k = 1$, i.e. $u = v$ a.e. in Ω, which is a contradiction. ∎

The proof of Proposition 3.1 follows the lines of the proof of Lemma 3.1 in LINDQVIST [17] for nondegenerate case ($a(x) \equiv 1$ in Ω).

Theorem 3.1 *The first eigenvalue $\lambda_1 > 0$ of the eigenvalue problem (3.4) is simple and there exists precisely one pair of normed eigenfunctions corresponding to λ_1 which do not change the sign in Ω.*

Proof Let u be the eigenfunction of (3.4) associated with $\lambda_1 > 0$ and changing sign in Ω. Then (3.5) holds with $\lambda = \lambda_1$ and test functions $\varphi = u^+$, $\varphi = u^-$. Hence, we get

$$\int a(x)|\nabla u^+|^p dx - \lambda_1 \int b(x)|u^+|^p dx = 0,$$

$$\int a(x)|\nabla u^-|^p dx - \lambda_1 \int b(x)|u^-|^p dx = 0.$$

It follows from here that both u^+ and u^- are nonnegative eigenfunctions associated with $\lambda_1 > 0$. Proposition 3.1 then implies that $u^+ = u^-$ a.e. in Ω which is a contradiction. The assertion follows, now, from Proposition 3.1. ∎

Remark 3.2 Let us remark that we have not proved the strict positivity of the first eigenfunction u_1 in Ω. This is the case when e.g. $a(x) \equiv 1$ (cf. ANANE [2], LINDQVIST [17]). The degeneracy (or the singularity) of $a(x)$ does not allow to apply the Harnack inequality in our case. For the same reason we cannot prove the local $C^{1,\alpha}$–regularity of the first eigenfunction as in the case $a(x) \equiv 1$ (cf. [17]). However, the results in this direction one can obtain when the degeneracy (or the singularity) of $a(x)$ is of the type "*power of the distance from the boundary $\partial\Omega$*". In this case we can restrict ourselves to any compact subset $\Omega' \subset\subset \Omega$ and apply Harnack type inequality of TRUDINGER [21] and regularity result of TOLKSDORF [20] to u_1 restricted to any Ω'. Thus we obtain that $u_1 > 0$ in Ω and locally $C^{1,\alpha}$ in Ω.

Lemma 3.3 *Let $J : W_0^{1,p}(w, \Omega) \longrightarrow [W_0^{1,p}(w, \Omega)]^*$ be an operator defined by*

$$\langle J(u), \varphi \rangle = \int_\Omega a(x)|\nabla u|^{p-2} \nabla u \nabla \varphi dx$$

for any $u, \varphi \in W_0^{1,p}(w, \Omega)$ (here $\langle \cdot, \cdot \rangle$ denotes the duality between $[W_0^{1,p}(w, \Omega)]^*$ and $W_0^{1,p}(w, \Omega)$). Then J is surjective and $J^{-1} : [W_0^{1,p}(w, \Omega)]^* \longrightarrow W_0^{1,p}(w, \Omega)$ is bounded and continuous.

Proof The operator J is bounded, strictly monotone, continuous and coercive. Then it follows from the Browder theorem (see e.g. FUČÍK, KUFNER [11]) that J is surjective. It follows from the Hölder inequality that

$$\langle J(v) - J(u), v - u \rangle \geq (\|v\|_a^{p-1} - \|u\|_a^{p-1})(\|v\|_a - \|u\|_a) \tag{3.27}$$

for any $u, v \in W_0^{1,p}(w, \Omega)$. The boundedness of J^{-1} follows immediately from (3.27). Let us suppose to the contrary that J^{-1} is not continuous. Then there exists a sequence (f_n) such that $f_n \to f$ in $[W_0^{1,p}(w, \Omega)]^*$ and $\|J^{-1}(f_n) - J^{-1}(f)\|_a \geq \delta$ for some $\delta > 0$. Denote $u_n = J^{-1}(f_n), u = J^{-1}(f)$. It follows from (3.27) that

$$\|f_n\|_* \cdot \|u_n\|_a \geq \langle f_n, u_n \rangle = \langle J(u_n), u_n \rangle \geq \|u_n\|_a^p,$$

i.e.

$$\|u_n\|_a^{p-1} \leq \|f_n\|_*$$

($\|\cdot\|_*$ denotes the norm in the dual space $[W_0^{1,p}(w, \Omega)]^*$). Then (u_n) is bounded in $W_0^{1,p}(w, \Omega)$ and we can assume that there exists $\tilde{u} \in W_0^{1,p}(w, \Omega)$ such that $u_n \rightharpoonup \tilde{u}$ in $W_0^{1,p}(w, \Omega)$. Hence we have

$$\begin{aligned}\langle J(u_n) - J(\tilde{u}), u_n - \tilde{u} \rangle &= \\ = \langle J(u_n) - J(u), u_n - \tilde{u} \rangle &+ \langle J(u) - J(\tilde{u}), u_n - \tilde{u} \rangle \longrightarrow 0\end{aligned} \tag{3.28}$$

since $J(u_n) \to J(u)$ in $[W_0^{1,p}(w, \Omega)]^*$. It follows from (3.27) (where we set $v = u_n$, $u = \tilde{u}$), (3.28) that $\|u_n\|_a \to \|\tilde{u}\|_a$. The uniform convexity of $W_0^{1,p}(w, \Omega)$ equipped with the norm $\|\cdot\|_a$ (see Section 2) implies $u_n \to \tilde{u}$ in $W_0^{1,p}(w, \Omega)$. This convergence together with the convergence $J(u_n) \to J(u)$ in $[W_0^{1,p}(w, \Omega)]^*$ imply $\tilde{u} = u$ which is a contradiction. The continuity of J^{-1} is proved. ■

4. Nonhomogeneous eigenvalue problem

In this section we will consider *nonhomogeneous eigenvalue problem*

$$\begin{aligned}-\text{div}(a(x, u)|\nabla u|^{p-2} \nabla u) &= \lambda b(x, u)|u|^{p-2}u \quad \text{in } \Omega, \\ u &= 0 \quad \text{on } \partial\Omega.\end{aligned} \tag{4.1}$$

Let $g : [0, \infty) \to [1, \infty)$ be a nondecreasing function, $\alpha(x) \in L^{\frac{q^*}{q^*-p}}(\Omega)$ for $q > q^* > p, \alpha(x) \in L^\infty(\Omega)$ for $q^* = p$ (for q, q^* see Section 3), $\beta > 0$ a constant. We assume that $a(x, s), b(x, s)$ are the Carathéodory functions (i.e. continuous in s for a.e. $x \in \Omega$ and measurable in x for all $s \in \mathbb{R}$) and,

$$\frac{w(x)}{c_1} \leq a(x, s) \leq c_1 g(|s|) w(x), \tag{4.2}$$

125

$$0 \leq b(x,s) \leq \alpha(x) + \beta |s|^{q^*-p} \tag{4.3}$$

hold for a.e. $x \in \Omega$ and for all $s \in \mathbb{R}$.

Moreover, assume that

$$\text{meas } \{x \in \Omega; b(x, v(x)) > 0\} > 0 \tag{4.4}$$

for any $v \in L^{q^*}(\Omega), v \not\equiv 0$. (Note that the condition (4.4) is fulfilled e.g. if $b(x,s) > 0$ for a.e. $x \in \Omega$ and for all $s \neq 0$.)

Definition 4.1 We will say that $\lambda \in \mathbb{R}$ is the *eigenvalue* and $u \in W_0^{1,p}(w, \Omega), u \not\equiv 0$, is the corresponding *eigenfunction* of the eigenvalue problem (4.1) if

$$\int_\Omega a(x, u(x)) |\nabla u|^{p-2} \nabla u \nabla \varphi \, dx = \lambda \int_\Omega b(x, u(x)) |u|^{p-2} u \varphi \, dx \tag{4.5}$$

holds for any $\varphi \in W_0^{1,p}(w, \Omega)$.

Proposition 4.1 *Let $u \in L^\infty(\Omega)$, $\|u\|_{q^*} = R > 0$, $u \geq 0$ be any eigenfunction of (4.1) corresponding to the eigenvalue λ. Then there exists $d(R) > 0$ (independent of g) such that $\|u\|_\infty \leq d(R)$.*

Proof Choose $\varphi = u^{\kappa p + 1}$ in (4.5) with $\kappa \geq 0$. We obtain

$$(\kappa p + 1) \int_\Omega a(x, u(x)) u^{\kappa p} |\nabla u|^p dx = \lambda \int_\Omega b(x, u(x)) u^{(\kappa+1)p} dx, \text{ i.e.}$$

$$\frac{\kappa p + 1}{(\kappa + 1)^p} \int_\Omega a(x, u(x)) |\nabla (u^{\kappa+1})|^p dx = \lambda \int_\Omega b(x, u(x)) u^{(\kappa+1)p} dx. \tag{4.6}$$

Now, the proof follows the lines of that of Lemma 3.2 using the assumptions (4.2), (4.3) instead of (3.1), (3.2), (3.3). ∎

We will define, now, equivalent problem to (4.1). Let $R > 0$ and $d = d(R) > 0$ be as above. We define

$$\tilde{a}(x, s) = \begin{cases} a(x, s) & \text{for } x \in \Omega, |s| \leq d(R), \\ a(x, d(R)) & \text{for } x \in \Omega, s > d(R), \\ a(x, -d(R)) & \text{for } x \in \Omega, s < -d(R). \end{cases} \tag{4.7}$$

Let us consider nonhomogeneous eigenvalue problem

$$\begin{aligned} -\text{div}(\tilde{a}(x,u)|\nabla u|^{p-2} \nabla u) &= \lambda b(x,u)|u|^{p-2} u \quad \text{in } \Omega, \\ u &= 0 \quad \text{on } \partial\Omega. \end{aligned} \tag{4.8}$$

Then it follows from Proposition 4.1 that $u \in W_0^{1,p}(w,\Omega), \|u\|_{q^*} = R, u \geq 0$ is the eigenfunction of (4.8) *if and only if* it is the eigenfunction of (4.1).

We will apply the Schauder fixed point theorem. For a given $v \in L^{q^*}(\Omega)$ set $a_v(x) = \tilde{a}(x, v(x)), b_v(x) = b(x, v(x))$. It follows from (4.2), (4.3), (4.4) and (4.7) that $a_v(x)$ and $b_v(x)$ fulfil (3.1), (3.2), (3.3) for any fixed $v \in L^{q^*}(\Omega)$. Let us consider homogeneous eigenvalue problem

$$-\operatorname{div}(a_v(x)|\nabla u|^{p-2}\nabla u) = \lambda b_v(x)|u|^{p-2}u \quad \text{in } \Omega,$$
$$u = 0 \quad \text{on } \partial\Omega \tag{4.9}$$

for any fixed $v \in L^{q^*}(\Omega)$. Due to the results of Section 3 there exists the *least* eigenvalue $\lambda_v > 0$ of (4.9) and *precisely one* corresponding eigenfunction u_v such that $u_v \geq 0$ a.e. in Ω, $u_v \in L^\infty(\Omega)$ and $\|u_v\|_{q^*} = R$. Hence we can define the *operator*

$$S : L^{q^*}(\Omega) \to L^{q^*}(\Omega)$$

which associates to $v \in L^{q^*}(\Omega)$ the first nonnegative eigenfunction u_v of (4.9) such that $\|u_v\|_{q^*} = R$.

Let us assume for a moment that S is a *compact operator*. Since it maps the ball $B_R = \{u \in L^{q^*}(\Omega), \|u\|_{q^*} \leq R\}$ into itself it follows from the Schauder fixed point theorem (see e.g. FUČÍK, KUFNER [11]) that S has a *fixed point* $u \in B_R$. Hence there exists $\lambda_u > 0$ such that

$$-\operatorname{div}(a_u(x)|\nabla u|^{p-2}\nabla u) = \lambda_u b_u(x)|u|^{p-2}u \quad \text{in } \Omega,$$
$$u = 0 \quad \text{on } \partial\Omega,$$

and it follows from the considerations above that $\lambda_u > 0$ is the least eigenvalue of (4.1) and $u \in L^\infty(\Omega), u \geq 0$ a.e. in Ω, is the corresponding eigenfunction satisfying $\|u\|_{q^*} = R$.

The *main result of this section* follows from the considerations presented above.

Theorem 4.1 *Let the assumptions from the beginning of Section 4 be fulfilled. Then for a given real number $R > 0$ there exists the least eigenvalue $\lambda > 0$ and the corresponding eigenfunction $u \in W_0^{1,p}(w,\Omega) \cap L^\infty(\Omega)$ of nonhomogeneous eigenvalue problem (4.1) such that $u \geq 0$ a.e. in Ω and $\|u\|_{q^*} = R$.*

In what follows it remains to prove *the compactness of the operator S*.
Let us define the Němytskii operators

$$G_1 : u \mapsto |u|^{p-2}u, \quad G_2 : u \mapsto |u|^p, \quad G_3 : u \mapsto b(x, u(x)).$$

Then G_i is bounded and continuous operator from $L^{q^*}(\Omega)$ into $L^{\frac{q^*}{p-1}}(\Omega)$ for $i = 1$, from $L^{q^*}(\Omega)$ into $L^{\frac{q^*}{p}}(\Omega)$ for $i = 2$, and from $L^{q^*}(\Omega)$ into $L^{\frac{q^*}{q^*-p}}(\Omega)$ for $i = 3$ (see e.g.

VAJNBERG [22], FUČÍK, KUFNER [11]). The Němytskii operator

$$G_4 : (u, z_1, \ldots, z_n) \mapsto \tilde{a}(x, u(x))(z_1^2(x) + \cdots + z_n^2(x))^{\frac{p-1}{2}}$$

is bounded and continuous from $L^{q^*}(\Omega) \times L^p(w, \Omega) \times \ldots \times L^p(w, \Omega)$ into $L^{\frac{p}{p-1}}(w^{-\frac{1}{p-1}}, \Omega)$ (see e.g. DRÁBEK, KUFNER, NICOLOSI [9], KUFNER, SÄNDIG [16]).

Lemma 4.1 Let $z, z_n \in W_0^{1,p}(w, \Omega)$ and

$$\int_\Omega a_v(x) |\nabla z|^{p-2} \nabla z \nabla \varphi dx = \int_\Omega f(x) \varphi(x) dx,$$

$$\int_\Omega a_{v_n}(x) |\nabla z_n|^{p-2} \nabla z_n \nabla \psi dx = \int_\Omega f_n(x) \psi(x) dx,$$

for any $\varphi, \psi \in W_0^{1,p}(w, \Omega)$ and let $v_n \to v$ in $L^{q^*}(\Omega), f_n \to f$ in $[W_0^{1,p}(w, \Omega)]^*$. Then $z_n \to z$ in $W_0^{1,p}(w, \Omega)$.

Proof Define the operators $J, J_n : W_0^{1,p}(w, \Omega) \to [W_0^{1,p}(w, \Omega)]^*$ by

$$\langle J(u), \varphi \rangle = \int_\Omega a_v(x) |\nabla u|^{p-2} \nabla u \nabla \varphi dx,$$

$$\langle J_n(u), \psi \rangle = \int_\Omega a_{v_n}(x) |\nabla u|^{p-2} \nabla u \nabla \psi dx,$$

for any $\varphi, \psi, u \in W_0^{1,p}(w, \Omega)$. Hence $J(z) = f$ and $J_n(z_n) = f_n$.
Let $n \in \mathbb{N}$ be fixed. Consider the equation

$$J_n(u) = h.$$

It follows from here

$$\int_\Omega a_{v_n}(x) |\nabla u|^p dx = \int_\Omega h(x) u(x) dx,$$

$$\|u\|_w^p \leq c_2 \|h\|_* \|u\|_w,$$

$$\|J_n^{-1}(h)\|_w \leq c_2 \|h\|_*^{\frac{1}{p-1}} \tag{1.10}$$

for any $h \in [W_0^{1,p}(w, \Omega)]^*$, where $c_2 > 0$ is independent of n and h. Analogously

$$\|J^{-1}(h)\|_w \leq c_2 \|h\|_*^{\frac{1}{p-1}} \tag{4.11}$$

(cf. Lemma 3.3). Applying Lemma 3.3 to $a(x) := a_v(x)$ we obtain continuity of J^{-1} (with J defined in this section).
Assume that (u_n) is the sequence satisfying $u_n \to z$ in $W_0^{1,p}(w, \Omega)$. It follows from the continuity of the Němytskii operator G_4 that

$$\|J_n(u_n) - J(u_n)\|_* = \sup_{\|\varphi\|_w \leq 1} |\langle J_n(u_n) - J(u_n), \varphi\rangle| =$$

$$= \sup_{\|\varphi\|_w \leq 1} |\int_\Omega (a_{v_n}(x) - a_v(x))| \nabla u_n|^{p-2} \nabla u_n \nabla \varphi dx| \leq$$

$$\leq \sup_{\|\varphi\|_w \leq 1} |\int_\Omega [a_{v_n}(x)| \nabla u_n|^{p-2} \nabla u_n - a_v(x)| \nabla z|^{p-2} \nabla z] \nabla \varphi dx| +$$

$$+ \sup_{\|\varphi\|_w \leq 1} |\int_\Omega [a_v(x)| \nabla z|^{p-2} \nabla z - a_v(x)| \nabla u_n|^{p-2} \nabla u_n] \nabla \varphi dx| \leq$$

(4.12)

$$\leq \sup_{\|\varphi\|_w \leq 1} \left(\int_\Omega w(x)^{-\frac{1}{p-1}} |a_{v_n}(x)| \nabla u_n|^{p-2} \nabla u_n - a_v(x)| \nabla z|^{p-2} \nabla z |^{\frac{p}{p-1}} dx \right)^{\frac{p-1}{p}}$$

$$\cdot \left(\int_\Omega w(x)|\nabla \varphi|^p dx \right)^{\frac{1}{p}} +$$

$$+ \sup_{\|\varphi\|_w \leq 1} \left(\int_\Omega w(x)^{-\frac{p-1}{p}} |a_v(x)| \nabla z|^{p-2} \nabla z - a_v(x)| \nabla u_n|^{p-2} \nabla u_n |^{\frac{p}{p-1}} dx \right)^{\frac{p-1}{p}}$$

$$\left(\int_\Omega w(x)|\nabla \varphi|^p dx \right)^{\frac{1}{p}} \to 0$$

for $n \to \infty$.

Set $u_n = J^{-1}(f_n)$. Then the assumptions of lemma and continuity of J^{-1} imply

$$u_n \to z \text{ in } W_0^{1,p}(w, \Omega). \tag{4.13}$$

The relations (4.10) – (4.13) and continuity of J^{-1} now yield

$$\|z_n - z\|_w \leq \|J_n^{-1}(f_n) - J^{-1}(f_n)\|_w + \|J^{-1}(f_n) - J^{-1}(f)\|_w \leq$$

$$\leq \|J_n^{-1}(J_n - J)J^{-1}(f_n)\|_w + \|J^{-1}(f_n) - J^{-1}(f)\|_w \leq$$

$$\leq c_2 \|J_n(u_n) - J(u_n)\|_*^{\frac{1}{p-1}} + \|J^{-1}(f_n) - J^{-1}(f)\|_w \to 0$$

for $n \to \infty$, which completes the proof. ∎

Proposition 4.2 *The operator $S : L^{q^*}(\Omega) \to L^{q^*}(\Omega)$ is compact.*

Proof We prove that S is continuous operator from $L^{q^*}(\Omega)$ into $W_0^{1,p}(w,\Omega)$. The assertion then follows from the compact imbedding $W_0^{1,p}(w,\Omega) \hookrightarrow\hookrightarrow L^{q^*}(\Omega)$ (see Section 2). Let $u_{v_n} = S(v_n), u_v = S(v)$. Suppose to the contrary that $v_n \to v$ in $L^{q^*}(\Omega)$ and

$$\|u_{v_n} - u_v\|_w \geq \delta \tag{4.14}$$

for some $\delta > 0$. We have

$$\int_\Omega a_v(x)|\nabla u_v|^{p-2} \nabla u_v \nabla \varphi dx = \lambda_v \int_\Omega b_v(x)|u_v|^{p-2} u_v \varphi dx, \tag{4.15}$$

$$\int_\Omega a_{v_n}(x)|\nabla u_{v_n}|^{p-2} \nabla u_{v_n} \nabla \psi dx = \lambda_{v_n} \int_\Omega b_{v_n}(x)|u_{v_n}|^{p-2} u_{v_n} \psi dx, \tag{4.16}$$

for any $\varphi, \psi \in W_0^{1,p}(w,\Omega)$. It follows from Lemma 3.3 that for any $v_n \in L^{q^*}(\Omega)$ there exists $z_n \in W_0^{1,p}(w,\Omega)$ such that

$$\int_\Omega a_{v_n}(x)|\nabla z_n|^{p-2} \nabla z_n \nabla \varphi dx = \lambda_v \int_\Omega b_v(x)|u_v|^{p-2} u_v \varphi dx, \tag{4.17}$$

for any $\varphi \in W_0^{1,p}(w,\Omega)$. Lemma 4.1 yields $z_n \to u_v$ in $W_0^{1,p}(w,\Omega)$ (and hence also in $L^{q^*}(\Omega)$). Applying the Hölder inequality, (4.3) and the Minkowski inequality, we obtain

$$|\int_\Omega b(x,v(x))|u_v|^{p-2} u_v(z_n - u_v) dx| \leq$$
$$\leq \left(\int_\Omega (b(x,v(x)))^{\frac{q^*}{q^*-1}} |u_v|^{\frac{q^*(p-1)}{q^*-1}} dx\right)^{\frac{q^*-1}{q^*}} \left(\int_\Omega |z_n - u_v|^{q^*} dx\right)^{\frac{1}{q^*}} \leq$$
$$\leq \left(\int_\Omega (b(x,v(x)))^{\frac{q^*}{q^*-p}} dx\right)^{\frac{q^*-p}{q^*}} \cdot$$
$$\cdot \left(\int_\Omega |u_v|^{q^*} dx\right)^{\frac{p-1}{q^*}} \left(\int_\Omega |z_n - u_v|^{q^*} dx\right)^{\frac{1}{q^*}} \leq \tag{4.18}$$
$$\leq \left[\left(\int_\Omega \alpha(x)^{\frac{q^*}{q^*-p}} p dx\right)^{\frac{q^*-p}{q^*}} + \beta\left(\int_\Omega |v(x)|^{q^*} dx\right)^{\frac{q^*-p}{q^*}}\right] \cdot$$
$$\cdot \left(\int_\Omega |u_v|^{q^*} dx\right)^{\frac{p-1}{q^*}} \left(\int_\Omega |z_n - u_v|^{q^*} dx\right)^{\frac{1}{q^*}} \to 0$$

for $n \to \infty$. Applying the Hölder inequality, (4.3), the Minkowski inequality and the continuity of the Němytskii operators G_2, G_3 we obtain

$$\left| \int_\Omega [b(x,v_n(x))|z_n|^p - b(x,v(x))|u_v|^p] dx \right| \le$$

$$\le \left| \int_\Omega b(x,v_n(x))[|z_n|^p - |u_v|^p] dx \right| +$$

$$+ \left| \int_\Omega [b(x,v_n(x)) - b(x,v(x))]|u_v|^p dx \right| \le$$

$$\le \left[\left(\int_\Omega \alpha(x)^{\frac{q^*}{q^*-p}} dx \right)^{\frac{q^*-p}{q^*}} + \beta \left(\int_\Omega |v_n(x)|^{q^*} dx \right)^{\frac{q^*-p}{q^*}} \right] \cdot \quad (4.19)$$

$$\cdot \left(\int_\Omega ||z_n|^p - |u_v|^p|^{\frac{q^*}{p}} dx \right)^{\frac{p}{q^*}} +$$

$$+ \left(\int_\Omega |b(x,v_n(x)) - b(x,v(x))|^{\frac{q^*}{q^*-p}} dx \right)^{\frac{q^*-p}{q^*}} \left(\int_\Omega |u_v|^{q^*} dx \right)^{\frac{p}{q^*}} \to 0$$

for $n \to \infty$. It follows from the variational characterization of λ_{v_n}, (4.15) – (4.19):

$$\lambda_{v_n} \le \frac{\int_\Omega a_{v_n}(x)|\nabla z_n|^p dx}{\int_\Omega b_{v_n}(x)|z_n|^p dx} = \frac{\lambda_v \int_\Omega b_v(x)|u_v|^{p-2} u_v z_n dx}{\int_\Omega b_{v_n}(x)|z_n|^p dx} \to$$

$$\to \lambda_v \frac{\int_\Omega b_v(x)|u_v|^p dx}{\int_\Omega b_v(x)|u_v|^p dx} = \lambda_v.$$

Hence

$$\limsup \lambda_{v_n} \le \lambda_v. \qquad (4.20)$$

Applying the Hölder inequality, the Minkowski inequality and the assumptions (4.2), (4.3) we obtain from (4.16) (with $\psi = u_{v_n}$):

$$\frac{1}{c_1} \|u_{v_n}\|_w^p \le \int_\Omega a_{v_n}(x)|\nabla u_{v_n}|^p dx = \lambda_{v_n} \int_\Omega b_{v_n}(x)|u_{v_n}|^p dx \le$$

$$\le \lambda_{v_n} \left[\left(\int_\Omega |\alpha(x)|^{\frac{q^*}{q^*-p}} dx \right)^{\frac{q^*-p}{q^*}} + \beta \left(\int_\Omega |v_n(x)|^{q^*} dx \right)^{\frac{q^*-p}{q^*}} \right] \left(\int_\Omega |u_{v_n}|^{q^*} dx \right)^{\frac{p}{q^*}}. \qquad (4.21)$$

It follows from the assumption $\|u_{v_n}\|_{q^*} = R$, from $v_n \to v$ in $L^{q^*}(\Omega)$ and from (4.21) that

$$\|u_{v_n}\|_w \le \text{const} \qquad (4.22)$$

for any $n \in \mathbb{N}$. Due to (4.22) we have

$$u_{v_n} \rightharpoonup u \text{ in } W_0^{1,p}(w,\Omega) \qquad (4.23)$$

(at least for some subsequence) for some $u \in W_0^{1,p}(w,\Omega)$ and hence $u_n \to u$ in $L^{q^*}(\Omega)$.

The Hölder inequality, the Minkowski inequality, (4.3) and the continuity of the Němytskii operators G_1 and G_3 imply

$$\left| \int_\Omega [b(x, v_n(x))|u_{v_n}|^{p-2}u_{v_n} - b(x, v(x))|u|^{p-2}u]\varphi dx \right| \leq$$

$$\leq \left| \int_\Omega [b(x, v_n(x)) - b(x, v(x))]|u_{v_n}|^{p-2}u_{v_n}\varphi dx \right| +$$

$$+ \left| \int_\Omega [b(x, v(x))[|u_{v_n}|^{p-2}u_{v_n} - |u|^{p-2}u]\varphi dx \right| \leq$$

$$\leq \left(\int_\Omega |b(x, v_n(x)) - b(x, v(x))|^{\frac{q^*}{q^*-p}} dx \right)^{\frac{q^*-p}{q^*}} \left(\int_\Omega |u_{v_n}|^{q^*} dx \right)^{\frac{p-1}{q^*}} \cdot \quad (4.24)$$

$$\cdot \left(\int_\Omega |\varphi|^{q^*} dx \right)^{\frac{1}{q^*}} + \left[\left(\int_\Omega |\alpha(x)|^{\frac{q^*}{q^*-p}} dx \right)^{\frac{q^*-p}{q^*}} + \beta \left(\int_\Omega |v(x)|^{\frac{q^*}{q^*-p}} dx \right)^{\frac{q^*-p}{q^*}} \right] \cdot$$

$$\cdot \left(\int_\Omega ||u_{v_n}|^{p-2}u_{v_n} - |u|^{p-2}u|^{\frac{q^*}{p-1}} dx \right)^{\frac{p-1}{q^*}} \left(\int_\Omega |\varphi|^{q^*} dx \right)^{\frac{1}{q^*}} \to 0.$$

for any $\varphi \in W_0^{1,p}(w, \Omega)$. Passing to suitable subsequences we can assume that

$$\lambda_{v_n} \to \lambda \in [0, \lambda_v] \quad (4.25)$$

(see (4.20)).

Let $\bar{u} \in W_0^{1,p}(w, \Omega)$ be the unique solution of

$$\int_\Omega a_v(x)|\nabla \bar{u}|^{p-2}\nabla \bar{u} \nabla \varphi dx = \lambda \int_\Omega b_v(x)|u|^{p-2}u\varphi dx \quad (4.26)$$

for any $\varphi \in W_0^{1,p}(w, \Omega)$ (Lemma 3.3 guarantees the existence of $\bar{u}$). It follows from (4.24) – (4.26) and from Lemma 4.1 that

$$u_{v_n} \to \bar{u} \text{ in } W_0^{1,p}(w, \Omega). \quad (4.27)$$

Now, (4.23), (4.27) imply $u = \bar{u}$ and $u_{v_n} \to u$ in $W_0^{1,p}(w, \Omega)$. Hence we have

$$\lambda_v \geq \lambda = \frac{\int_\Omega a_v(x)|\nabla u|^p dx}{\int_\Omega b_v(x)|u|^p dx} \geq \inf_{\substack{\tilde{u} \neq 0 \\ \tilde{u} \in W_0^{1,p}(w,\Omega)}} \frac{\int_\Omega a_v(x)|\nabla \tilde{u}|^p dx}{\int_\Omega b_v(x)|\tilde{u}|^p dx} =$$

$$= \frac{\int_\Omega a_v(x)|\nabla u_v|^p dx}{\int_\Omega b_v(x)|u_v|^p dx} = \lambda_v.$$

It follows from here $\lambda = \lambda_v$ and $u = u_v$ (see the uniqueness of $u_v \geq 0$, $\|u_v\|_{q^*} = R$ in Section 3).

In particular, this means that

$$u_{v_n} \to u_v \text{ in } W_0^{1,p}(w, \Omega),$$

which contradicts (4.14). This completes the proof of Proposition 4.2. ∎

Remark 4.1 The proofs in Section 4 can be performed in the same way working with $L^\infty(\Omega)$ instead of $L^{\frac{q^*}{q^*-p}}(\Omega)$ in the case $q^* = p$. Hence we obtain the following *special version* of Theorem 4.1.

Theorem 4.2 *Let (4.2) – (4.4) be fulfilled with $a(x) \in L^\infty(\Omega)$ and $q^* = p$. Then for a given real number $R > 0$ there exists the least eigenvalue $\lambda > 0$ and the corresponding eigenfunction $u \in W_0^{1,p}(w, \Omega) \cap L^\infty(\Omega)$ of (4.1) such that $u \geq 0$ a.e. in Ω and $\|u\|_p = R$.*

Remark 4.2 Since the eigenvalue problem (4.9) is homogeneous, we can define the operator $\tilde{S} : L^{q^*}(\Omega) \longrightarrow L^{q^*}(\Omega)$ which associates to $v \in L^{q^*}(\Omega)$ the first nonpositive eigenfunction $-u_v$ of (4.9) such that $\| - u_v\|_{q^*} = R$. It is clear from the above considerations that $\tilde{S}$ has the *same properties* as S. Hence repeating the same arguments we prove the following dual version of Theorem 4.1.

Theorem 4.3 *Let the assumptions of Theorem 4.1 be fulfilled. Then for a given real number $R > 0$ there exists the least eigenvalue $\tilde{\lambda} > 0$ and the corresponding eigenfunction $\tilde{u} \in W_0^{1,p}(w, \Omega) \cap L^\infty(\Omega)$ of nonhomogeneous eigenvalue problem (4.1) such that $\tilde{u} \leq 0$ a.e. in Ω and $\|\tilde{u}\|_{q^*} = R$.*

Remark 4.3 Let λ and $\tilde{\lambda}$ be the least eigenvalue guaranteed by Theorem 4.1 and 4.3, respectively, for a given fixed $R > 0$. Then it may be $\lambda \neq \tilde{\lambda}$ due to the fact that the eigenvalue problem (4.1) is not homogeneous in general.

Remark 4.4 If the weight function w is of "the power type" in the sense of Remark 3.2, the eigenfunction in Theorem 4.1 (and 4.2) is *positive* in Ω and locally $C^{1,\alpha}$ in Ω.

Example 4.1 Let Ω be a bounded domain in $\mathbb{R}^n, p > 1, w(x)$ be positive and measurable in Ω satisfying $w(x) \in L_{loc}^1(\Omega)$, $\frac{1}{w(x)} \in L^s(\Omega)$ for $s > \max\{\frac{n}{p}, \frac{1}{p-1}\}$. Consider the eigenvalue problem

$$-\text{div}(w(x)e^{u^2}|\nabla u|^{p-2}\nabla u) = \lambda |u|^{p-2}u \quad \text{in } \Omega, \qquad (4.28)$$
$$u = 0 \quad \text{on } \partial\Omega.$$

In this case we have
$$a(x, s) = w(x)e^{s^2}, b(x, s) \equiv 1$$
for a.e. $x \in \Omega$ and for all $s \in \mathbb{R}$.

It follows from Theorem 4.2 that *for any given real number $R > 0$ there exists the least eigenvalue $\lambda > 0$ and the corresponding eigenfunction $u \in W_0^{1,p}(w, \Omega) \cap L^\infty(\Omega)$ of (4.28) such that $u \geq 0$ a.e. in Ω and $\|u\|_p = R$.* □

Example 4.2 Let us consider for Ω the plane domain $\Omega = (-1,1) \times (-1,1)$ (i.e. $\Omega \subset \mathbb{R}^2$). For $x = (x_1, x_2) \in \Omega$ set

$$w(x) = \begin{cases} 1, & x_1 \leq 0, \\ x_2^\nu (1-x_1)^\gamma, & x_1 > 0, x_2 > 0, \\ |x_2|^\mu (1-x_1)^\gamma, & x_1 > 0, x_2 < 0 \end{cases}$$

with ν, μ, γ real numbers. Consider the eigenvalue problem

$$\begin{aligned} -\mathrm{div}(w(x)(1+u^4)|\nabla u|^2 \nabla u) &= \lambda u^9 \quad \text{in } \Omega, \\ u &= 0 \quad \text{on } \partial\Omega. \end{aligned} \qquad (4.29)$$

In this case we have $p = 4$,

$$a(x,s) = w(x)(1+s^4), b(x,s) = s^6$$

for a.e. $x \in \Omega$ and for all $s \in \mathbb{R}$. Thus the principal part of the differential operator has a *degeneration* (or *singularity*) which is concentrated on a part Γ_1 of the boundary $\partial\Omega$,

$$\Gamma_1 = \{x = (x_1, x_2); x_1 = 1, x_2 \in (-1,1)\},$$

as well as on a segment Γ_2 in the interior of Ω,

$$\Gamma_2 = \{x = (x_1, x_2); x_1 \in (0,1), x_2 = 0\}.$$

Condition (2.1) indicates that we have to choose ν and μ from the interval $(-1,3)$ with no condition on γ. Let us assume that

$$\nu, \mu \in \left(-1, \frac{4}{3}\right), \quad \gamma \in \left(-\infty, \frac{4}{3}\right). \qquad (4.30)$$

It follows from (4.30) that $\frac{1}{w(x)} \in L^{\frac{3}{4}}(\Omega)$ and $q = 12$ (see Section 2). Hence the growth condition (4.3) is fulfilled e.g. with $q^* = 10$. Applying Theorem 4.1 and Remark 4.4 we have the following assertion.

Let us assume (4.30). Then for a given real number $R > 0$ there exists the least eigenvalue $\lambda > 0$ and the corresponding eigenfunction $u \in W_0^{1,4}(w, \Omega) \cap L^\infty(\Omega)$ of (4.29) such that $u > 0$ in Ω and $\|u\|_{10} = R$. Moreover, $u \in C^{1,\alpha}_{loc}(\Omega)$.

Note that for ν, μ and γ *positive* we have a *degeneration* of the same extent at Γ_1 and Γ_2. On the other hand, the *singularity* can occur in a limited extent at Γ_2 (for ν or μ negative, but bigger than -1), but big enough at Γ_1 (for any $\gamma < 0$). $\square$

5. Maximum principle for degenerate (singular) equations

In this and in the following section we will assume, for simplicity, that $q^* = p$, i.e. (4.2) – (4.4) are fulfilled with $\alpha(x) \in L^\infty(\Omega)$, $\beta = 0$ (cf. Remark 4.1).

Due to Theorem 4.1 we have a function
$$R \mapsto \lambda_R$$
mapping $(0,\infty)$ into $(0,\infty)$. Using the assumptions (4.2) and (4.3) we prove the estimate of λ_R from below uniformly with respect to $R > 0$. Set
$$\lambda^{\#} = \frac{1}{c_1 c_2 \|\alpha(x)\|_\infty},$$
where $c_2 > 0$ is the constant of the imbedding $W_0^{1,p}(w,\Omega) \hookrightarrow L^p(\Omega)$ i.e.
$$\|u\|_p \leq c_2 \|u\|_w \tag{5.1}$$
for any $u \in W_0^{1,p}(w,\Omega)$. Let us assume that $w \in W_0^{1,p}(w,\Omega)$ is such that both integrals
$$\int_\Omega a(x,v(x))|\nabla v|^p dx \text{ and } \int_\Omega b(x,v(x))|v|^p dx$$
are finite.

Lemma 5.1 *Let $\lambda \leq \lambda^{\#}$. Then*
$$\int_\Omega a(x,v(x))|\nabla v|^p dx - \lambda \int_\Omega b(x,v(x))|v|^p dx \geq 0. \tag{5.2}$$

Proof The assertion is clear if $\lambda \leq 0$. Let $\lambda > 0$. It follows from (4.2) that
$$\int_\Omega a(x,v(x))|\nabla v|^p dx \geq \frac{1}{c_1} \int_\Omega w(x)|\nabla v|^p dx = \frac{1}{c_1}\|v\|_w^p. \tag{5.3}$$
On the other hand applying (4.3) and (5.1) we obtain
$$\int_\Omega b(x,v(x))|v|^p dx \leq \|\alpha(x)\|_\infty \int_\Omega |v|^p dx \leq c_2 \|\alpha(x)\|_\infty \|v\|_w^p. \tag{5.4}$$
Combining (5.3) and (5.4) we get
$$\int_\Omega a(x,v(x))|\nabla v|^p dx - \lambda \int_\Omega b(x,v(x))|v|^p dx \geq \left[\frac{1}{c_1} - \lambda c_2 \|\alpha(x)\|_\infty\right] \|v\|_w^p. \tag{5.5}$$
Since $\lambda \leq \lambda^{\#}$ implies $\frac{1}{c_1} - \lambda c_2\|\alpha(x)\|_\infty \geq 0$ the assertion follows from (5.5). ∎

Lemma 5.2 *Let $\lambda^* = \inf_{R>0} \lambda_R$. Then $\lambda^* \geq \lambda^{\#}$.*

Proof Let $R > 0$ be arbitrary. Let λ_R be the eigenvalue and $u_R \in W_0^{1,p}(w, \Omega) \cap L^\infty(\Omega), \|u_R\|_p = R$ be the corresponding eigenfunction of (4.1). Choosing $\varphi = u_R$ as a test function in (4.5) we obtain

$$\int_\Omega a(x, u_R(x))|\nabla u_R|^p dx = \lambda_R \int_\Omega b(x, u_R(x))|u_R|^p dx. \qquad (5.6)$$

The estimates (5.3) and (5.4) with v replaced by u_R yield

$$\frac{1}{c_1} \leq \lambda_R c_2 \|\alpha(x)\|_\infty, \text{ i.e. } \lambda_R \geq \frac{1}{c_1 c_2 \|\alpha(x)\|_\infty} = \lambda^\#.$$

Since $R > 0$ is arbitrary we have $\lambda^* \geq \lambda$.

∎

Let us consider BVP

$$\begin{aligned}-\text{div}(a(x,u)|\nabla u|^{p-2}\nabla u) &= \vartheta b(x,u)|u|^{p-2}u + h \text{ in } \Omega, \\ u &= 0 \qquad\qquad\qquad\qquad \text{on } \partial\Omega, \end{aligned} \qquad (5.7)$$

with a real parameter ϑ and with $h(x) \in L^{p'}(\Omega), \frac{1}{p} + \frac{1}{p'} = 1$.

Definition 5.1 We will say that the BVP (5.7) satisfies *the maximum principle* if $h \geq 0$ in Ω implies $u \geq 0$ a.e. in Ω for all possible weak solutions $u \in W_0^{1,p}(w, \Omega)$ of (5.7).

Lemma 5.3 (Sufficient condition.) *Let* $b(x, s) > 0$ *for a.e.* $x \in \Omega$ *and for all* $s \in \mathbb{R}$. *If*

$$\vartheta < \lambda^\# \qquad (5.8)$$

then the BVP (5.7) satisfies the maximum principle.

Proof Let $u \in W_0^{1,p}(w, \Omega)$ be the weak solution of (5.7) corresponding to $h \in L^{p'}(\Omega)$, $h \geq 0$ a.e. in Ω. Then

$$\int_\Omega a(x,u)|\nabla u|^{p-2}\nabla u \nabla \varphi dx = \vartheta \int_\Omega b(x,u)|u|^{p-2}u\varphi dx + \int_\Omega h\varphi dx \qquad (5.9)$$

holds for any $\varphi \in W_0^{1,p}(w, \Omega)$. Choose $\varphi = u^- = \max\{-u(x), 0\}$ as a test function in (5.9). We obtain

$$-\int_\Omega a(x,u)|\nabla u^-|^p dx = -\vartheta \int_\Omega b(x,u)|u^-|^p dx + \int_\Omega hu^- dx. \qquad (5.10)$$

It follows from (5.10) and from the fact $h \geq 0$ a.e. in Ω that

$$\int_\Omega a(x,u^-)|\nabla u^-|^p dx - \vartheta \int_\Omega b(x,u^-)|u^-|^p dx \leq 0. \qquad (5.11)$$

On the other hand it follows from Lemma 5.1 (see (5.2)) that

$$\int_\Omega a(x,u^-)|\nabla u^-|^p dx - \lambda^\# \int_\Omega b(x,u^-)|u^-|^p dx \geq 0. \tag{5.12}$$

The inequalities (5.11) and (5.12) imply

$$(\lambda^\# - \vartheta) \int_\Omega b(x,u^-)|u^-|^p dx \leq 0.$$

Due to (5.8) and the assumption on $b(x,s)$ we get $u^- = 0$ a.e. in Ω, i.e. $u \geq 0$ a.e. in Ω. Hence the BVP (5.7) satisfies the maximum principle. ∎

Lemma 5.4 (Necessary condition.) *Let both $a(x,s)$ and $b(x,s)$ be even in s, i.e.*

$$a(x,s) = a(x,-s) \quad \text{and} \quad b(x,s) = b(x,-s)$$

for all $s \in \mathbb{R}$ and for a.e. $x \in \Omega$. Let the BVP (5.7) satisfy the maximum principle. Then

$$\vartheta < \lambda_R$$

for any $R > 0$.

Proof We will proceed via contradiction. Let us assume that there exists $R > 0$ such that $\vartheta \geq \lambda_R$. Let u_R be the corresponding nonnegative eigenfunction of (4.2) satisfying $\|u_R\|_p = R$ (see Theorem 4.1). Set

$$h(x) = (\vartheta - \lambda_R) b(x, u_R(x))|u_R(x)|^{p-2} u_R(x). \tag{5.13}$$

Then $h(x) \geq 0$ a.e. in Ω. Since both $a(x,s)$ and $b(x,s)$ are even in s, the function $-u_R$ is the solution of the BVP (5.7) with h given by (5.13). But $u_R \not\equiv 0$ and $-u_R \leq 0$ a.e. in Ω. This is a contradiction with the fact that the BVP (5.7) satisfies the maximum principle. ∎

Remark 5.1 Let us consider the case of p-Laplacian:

$$a(x,s) \equiv 1 \quad \text{and} \quad b(x,s) \equiv 1. \tag{5.14}$$

Then

$$\lambda_R \equiv \lambda^* = \lambda^\# = \lambda_1 \quad \text{for} \quad R > 0,$$

where $\lambda_1 > 0$ is the first eigenvalue of p-Laplacian (see DRÁBEK [7] and the references therein). The functions $a(x,s), b(x,s)$ given by (5.14) satisfy the assumptions of Lemmas 5.3 and 5.4. It follows from here that the BVP

$$-\text{div}(|\nabla u|^{p-2}\nabla u) = \vartheta|u|^{p-2}u + h \quad \text{in } \Omega,$$
$$u = 0 \quad \text{on } \partial\Omega,$$

satisfies the maximum principle *if and only if*

$$\vartheta < \lambda_1$$

(cf. FLECKINGER, HERNÁNDEZ, de THÉLIN [10]).

6. Solvability of degenerate (singular) BVP

In this section we will formulate and prove the existence results for the BVP (1.6). We will assume that $f(x,s)$ is the Carathéodory function satisfying the assumption

$$|f(x,s)| \leq \beta(x)|s|^{p-1} + \gamma(x), \tag{6.1}$$

for a.e. $x \in \Omega$ and for all $s \in \mathbb{R}$, where $\beta(x), \gamma(x) \in L^\infty(\Omega)$.

Definition 6.1 We will say that $u \in W_0^{1,p}(w,\Omega)$ is *the weak solution* of the BVP (1.6) if

$$\int_\Omega a(x,u(x))|\nabla u|^{p-2}\nabla u \nabla \varphi \, dx = \\ = \lambda \int_\Omega b(x,u(x))|u|^{p-2}u\varphi \, dx + \int_\Omega f(x,u(x))\varphi \, dx \tag{6.2}$$

holds for any $\varphi \in W_0^{1,p}(w,\Omega)$.

Theorem 6.1 *Let us assume (2.6), (4.2), (4.3) (with $q^* = p$, $\alpha \in L^\infty(\Omega)$, $\beta \equiv 0$), (6.1) and*

$$\lambda < \frac{1 - c_1 c_2 \|\beta(x)\|_\infty}{c_1 c_2 \|\alpha(x)\|_\infty} (= \lambda(1 - c_1 c_2 \|\beta\|_\infty)). \tag{6.3}$$

Then the BVP (1.6) has at least one weak solution $u \in W_0^{1,p}(w,\Omega) \cap L^\infty(\Omega)$.

Proof We will prove the assertion in three steps. In the first step we prove an apriori estimate in $L^r(\Omega)$ for any $1 \leq r \leq \infty$, for any possible solution of the BVP (1.6). In the second step we find a suitable operator representation of the BVP (1.6). The third step is devoted to the application of the degree theory in order to prove the existence of the weak solution.

Step 1 (apriori estimate in $L^r(\Omega), 1 \le r \le \infty$). Let us suppose that $u \in W_0^{1,p}(w,\Omega)$ is the weak solution of the BVP (1.6). For $\varphi \in W_0^{1,p}(w,\Omega)$ set formally

$$\langle T(u), \varphi \rangle = \int_\Omega a(x,u)|\nabla u|^{p-2}\nabla u \nabla \varphi \, dx - \lambda \int_\Omega b(x,u)|u|^{p-2}u\varphi \, dx -$$

$$- \int_\Omega f(x,u)\varphi \, dx.$$

Choosing $\varphi = u$ as a test function in (6.2) we obtain due to (4.2), (4.3), (6.1),(5.1),(5.2) that

$$0 = \langle T(u), u \rangle = \frac{\lambda}{\lambda^\#}[\int_\Omega a(x,u)|\nabla u|^p dx - \lambda^\# \int_\Omega b(x,u)|u|^p dx] +$$

$$+ \frac{\lambda^\# - \lambda}{\lambda^\#} \int_\Omega a(x,u)|\nabla u|^p dx - \int_\Omega f(x,u)u \, dx \ge$$

$$\ge \frac{\lambda^\# - \lambda}{c_1 \lambda^\#} \|u\|_w^p - c_2 \|\beta(x)\|_\infty \|u\|_w^p - \|\gamma(x)\|_\infty \|u\|_1 \ge$$

$$\ge \frac{\lambda^\#(1 - c_1 c_2 \|\beta(x)\|_\infty) - \lambda}{c_1 \lambda^\#} \|u\|_w^p - c_2(\operatorname{meas}\Omega)^{\frac{p-1}{p}} \|\gamma(x)\|_\infty \|u\|_w.$$
(6.4)

It follows from (6.3) and (6.4) that $\|u\|_w \le c_3$ for any possible weak solution u of the BVP (1.6). By the imbedding (2.4) we have $\|u\|_q \le c_4$. In particular, we also have

$$\|u^+\|_q \le c_4 \tag{6.5}$$

(here $u^+ = \max\{u(x), 0\}$). For a given real $M > 0$ set

$$v_M(x) = \inf\{u(x), M\} \quad \text{on} \quad \{x \in \Omega; u(x) \ge 0\},$$
$$v_M(x) = 0 \quad \text{on} \quad \{x \in \Omega, u(x) < 0\}.$$

For a real $\kappa \ge 0$ set $\varphi = v_M^{\kappa p+1} \in W_0^{1,p}(w,\Omega) \cap L^\infty(\Omega)$ and choose this φ as a test function in (6.2). We obtain

$$(\kappa p + 1) \int_{\Omega(u>0)} a(x,u)v_M^{\kappa p} |\nabla v_M|^p dx =$$

$$= \lambda \int_{\Omega(u>0)} b(x,u) u^{p-1} v_M^{\kappa p+1} dx + \int_{\Omega(u>0)} f(x,u) v_M^{\kappa p+1} dx.$$
(6.6)

The left hand side of (6.6) is estimated by using (2.4) and (4.2):

$$(\kappa p + 1) \int_{\Omega(u>0)} a(x,u) v_M^{\kappa p} |\nabla v_M|^p dx \geq$$
$$\geq \frac{\kappa p + 1}{c_1} \int_{\Omega(u>0)} w(x) v_M^{\kappa p} |\nabla v_M|^p dx =$$
$$= \frac{\kappa p + 1}{c_1 (\kappa + 1)^p} \int_{\Omega(u>0)} w(x) |\nabla (v_M^{\kappa+1})|^p dx \geq \qquad (6.7)$$
$$\geq \frac{\kappa p + 1}{c_1 c_2^p (\kappa + 1)^p} \Big(\int_{\Omega(u>0)} (v_M^{\kappa+1})^q dx \Big)^{\frac{p}{q}}.$$

The right hand side of (6.6) is estimated by using (4.3), (6.1) and the Hölder inequality:

$$\lambda \int_{\Omega(u>0)} b(x,u) u^{p-1} v_M^{\kappa p+1} dx + \int_{\Omega(u>0)} f(x,u) v_M^{(\kappa p+1)} dx \leq$$
$$\leq \lambda \|\alpha(x)\|_\infty \int_{\Omega(u>0)} u^{(\kappa+1)p} dx + \|\beta(x)\|_\infty \int_{\Omega(u>0)} u^{(\kappa+1)p} dx + \qquad (6.8)$$
$$+ \|\gamma(x)\|_\infty (\operatorname{meas} \Omega)^{\frac{p-1}{(\kappa+1)p}} \Big(\int_{\Omega(u>0)} u^{(\kappa+1)p} dx \Big)^{\frac{\kappa p+1}{(\kappa+1)p}}.$$

It follows from (6.6) – (6.8) that

$$\Big(\int_{\Omega(u>0)} v_M^{(\kappa+1)q} dx \Big)^{\frac{p}{q}} \leq c_5 \frac{(\kappa+1)^p}{\kappa p + 1} \Big[\int_{\Omega(u>0)} u^{(\kappa+1)p} dx + c_5' \Big]$$

with some $c_5, c_5' > 0$.

Now, we use (6.5) and argue in the same way as in the proof of Lemma 3.2 in order to get

$$\|u^+\|_r \leq c_6$$

for any $1 \leq r \leq \infty$ with a constant $c_6 > 0$ independent of r.

Similarly we can handle u^- and hence we have, finally, a priori estimate in $L^\infty(\Omega)$ of any possible solution of BVP (1.6):

$$\|u\|_\infty \leq d \qquad (6.9)$$

with some $d > 0$.

Step 2 (operator representation of the BVP (1.6)). Let us define the function

$$\tilde{a}(x,s) = \begin{cases} a(x,s) & \text{for } x \in \Omega, |s| \leq d, \\ a(x,d) & \text{for } x \in \Omega, s > d, \\ a(x,-d) & \text{for } x \in \Omega, s < -d. \end{cases} \qquad (6.10)$$

where $d > 0$ is the real number from (6.9). Then due to Step 1 the function $u \in W_0^{1,p}(w, \Omega)$ is the weak solution of the BVP (1.6) if and only if u satisfies (6.9) and it is the weak solution of the BVP

$$-\text{div}(\tilde{a}(x, u)|\nabla u|^{p-2}\nabla u) = \lambda b(x, u)|u|^{p-2}u + f(x, u) \text{ in } \Omega,$$
$$u = 0 \qquad \text{on } \partial\Omega. \tag{6.11}$$

Define the operators

$$A, B, F : W_0^{1,p}(w, \Omega) \to [W_0^{1,p}(w, \Omega)]^*$$

by the following way

$$\langle A(u), \varphi \rangle = \int_\Omega \tilde{a}(x, u)|\nabla u|^{p-2} \nabla u \nabla \varphi \, dx,$$

$$\langle B(u), \varphi \rangle = \int_\Omega b(x, u)|u|^{p-2} u \varphi \, dx,$$

$$\langle F(u), \varphi \rangle = \int_\Omega f(x, u) \varphi \, dx$$

for any $u, \varphi \in W_0^{1,p}(w, \Omega)$. Then (4.2), (4.3), (6.1) and (6.10) guarantee that A, B and F are well defined operators. Set

$$\tilde{T}(u) = A(u) - \lambda B(u) - F(u)$$

for any $u \in W_0^{1,p}(w, \Omega)$. It is easy to see that $u \in W_0^{1,p}(w, \Omega)$ is the weak solution of the BVP (6.11) if and only if u is the solution of the *operator equation*

$$\tilde{T}(u) = 0. \tag{6.12}$$

Step 3 (application of the degree theory). The operator

$$\tilde{T} : W_0^{1,p}(w, \Omega) \to [W_0^{1,p}(w, \Omega)]^*$$

defined in Step 2 satisfies the condition $\alpha(W_0^{1,p}(w, \Omega))$ (see SKRYPNIK [19], cf. the condition (S_+) in BROWDER, PETRYSHIN [6]): *Let u_n converge weakly to u_0 in $W_0^{1,p}(w, \Omega)$ and*

$$\limsup_{n \to \infty} \langle \tilde{T}(u_n), u_n - u_0 \rangle \leq 0 \tag{6.13}$$

hold. Then u_n converge strongly to u_0 in $W_0^{1,p}(w, \Omega)$.

Let us verify this condition. The weak convergence of u_n to u_0 in $W_0^{1,p}(w, \Omega)$ together with (2.5) imply the strong convergence of u_n to u_0 in $L^p(\Omega)$. Hence applying the Hölder inequality we obtain

$$|\langle B(u_n), u_n - u_0\rangle| = \left|\int_\Omega b(x, u_n)|u_n|^{p-2}u_n(u_n - u_0)dx\right| \le$$

$$\le \|\alpha(x)\|_\infty \left(\int_\Omega |u_n|^p dx\right)^{\frac{p-1}{p}} \left(\int_\Omega |u_n - u_0|^p dx\right)^{\frac{1}{p}} \le \quad (6.14)$$

$$\le c_7\|\alpha(x)\|_\infty \|u_n - u_0\|_p \to 0,$$

and

$$|\langle F(u_n), u_n - u_0\rangle| = \left|\int_\Omega f(x, u_n)(u_n - u_0)dx\right| \le$$

$$\le \|\beta(x)\|_\infty \left(\int_\Omega |u_n|^p dx\right)^{\frac{p-1}{p}} \left(\int_\Omega |u_n - u_0|^p dx\right)^{\frac{1}{p}} +$$

$$+ \|\gamma(x)\|_\infty \int_\Omega |u_n - u_0|dx \le \quad (6.15)$$

$$\le c_7\|\beta(x)\|_\infty \|u_n - u_0\|_p + c_8\|\gamma(x)\|_\infty \|u_n - u_0\|_p \to 0$$

for $n \to \infty$. It follows from (6.13) - (6.15) that

$$0 \ge \limsup_{n \to \infty} \langle \tilde{T}(u_n), u_n - u_0\rangle =$$

$$= \limsup_{n \to \infty} \langle A(u_n) - \lambda B(u_n) - F(u_n), u_n - u_0\rangle = \quad (6.16)$$

$$= \limsup_{n \to \infty} \langle A(u_n), u_n - u_0\rangle = \limsup_{n \to \infty} \langle A(u_n) - A(u_0), u_n - u_0\rangle.$$

The continuity of the Nemytskii operator

$$G : (u, z_1, \cdots, z_n) \mapsto \tilde{a}(x, u)(z_1^2 + \cdots + z_n^2)^{\frac{p-1}{2}}$$

from $L^p(\Omega) \times L^p(w, \Omega) \times \cdots \times L^p(w, \Omega)$ into $L^{\frac{p}{p-1}}(w^{-\frac{1}{p-1}}, \Omega)$, the boundedness of u_n in $W_0^{1,p}(w, \Omega)$ and the Hölder inequality imply

$$\int_\Omega [\tilde{a}(x, u_n) - \tilde{a}(x, u_0)]|\nabla u_0|^{p-2}\nabla u_0(\nabla u_n - \nabla u_0)dx \le$$

$$\le \left(\int_\Omega w^{-\frac{1}{p-1}}|\tilde{a}(x, u_n)|\nabla u_0|^{p-2}\nabla u_0 - \tilde{a}(x, u_0)|\nabla u_0|^{p-2}\nabla u_0|^{\frac{p}{p-1}}dx\right)^{\frac{p-1}{p}} \cdot$$

$$\cdot \left(\int_\Omega w|\nabla u_n - \nabla u_0|^p dx\right)^{\frac{1}{p}} \le \quad (6.17)$$

$$\le c_9\|G(u_n, \nabla u_0) - G(u_0, \nabla u_0)\|_{L^{\frac{p}{p-1}}(w^{-\frac{1}{p-1}}, \Omega)} \to 0$$

for $n \to \infty$. It follows from the Hölder inequality, (4.2), (6.16) and (6.17) that

$$0 \geq \limsup_{n \to \infty} \langle A(u_n) - A(u_0), u_n - u_0 \rangle =$$

$$= \limsup_{n \to \infty} \left\{ \int_\Omega [\tilde{a}(x, u_n) - \tilde{a}(x, u_0)] |\nabla u_0|^{p-2} \nabla u_0 (\nabla u_n - \nabla u_0) dx \right.$$

$$\left. + \int_\Omega \tilde{a}(x, u_n)[|\nabla u_n|^{p-2} \nabla u_n - |\nabla u_0|^{p-2} \nabla u_0](\nabla u_n - \nabla u_0) dx \right\} \geq \quad (6.18)$$

$$\geq \frac{1}{c_1} \limsup_{n \to \infty} \int_\Omega w[|\nabla u_n|^{p-2} \nabla u_n - |\nabla u_0|^{p-2} \nabla u_0](\nabla u_n - \nabla u_0) dx \geq$$

$$\geq \frac{1}{c_1} \limsup_{n \to \infty} (\|u_n\|_w^{p-1} - \|u_0\|_w^{p-1})(\|u_n\|_w - \|u_0\|_w).$$

We obtain from (6.18) that

$$\|u_n\|_w \to \|u_0\|_w$$

for $n \to \infty$. The uniform convexity of $W_0^{1,p}(w, \Omega)$ then implies that u_n converge strongly to u_0 in $W_0^{1,p}(w, \Omega)$. Thus the condition $\alpha(W_0^{1,p}(w, \Omega))$ is satisfied.

Similarly as in (6.4) we prove that

$$\langle \tilde{T}(u), u \rangle > 0 \quad (6.19)$$

for any $u \in \partial B_\rho(0) = \{u \in W_0^{1,p}(w, \Omega); \|u\|_w = \rho\}$ with $\rho > 0$ sufficiently large (it is sufficient to verify that $\tilde{a}(x, s)$ satisfies the same bound from below as $a(x, s)$). Hence the degree deg $[\tilde{T}; B_\rho(0), 0]$ is well defined (see SKRYPNIK [19]) and due to (6.19) we have

$$\deg [\tilde{T}; B_\rho(0), 0] = 1. \quad (6.20)$$

The basic property of the degree and (6.20) yield that there is $u \in W_0^{1,p}(w, \Omega)$, $\|u\|_w < \rho$ satisfying (6.12). Due to the discussions in Steps 1 and 2 we have $u \in L^\infty(\Omega)$ and it is the weak solution of the BVP (1.6).

∎

Theorem 6.2 *Let us assume the same as in Theorem 6.1 and, moreover, suppose that $b(x, s) > 0, f(x, s) \geq 0$ for a.e. $x \in \Omega$ and for all $s \in \mathbb{R}$. Then the BVP (1.6) has at least one weak solution $u \in W_0^{1,p}(w, \Omega) \cap L^\infty(\Omega)$ satisfying $u \geq 0$ a.e. in Ω.*

Proof It follows from Theorem 6.1 that the BVP (1.6) has at least one weak solution $u \in W_0^{1,p}(w, \Omega) \cap L^\infty(\Omega)$. Let us apply Lemma 5.3 where we put $\vartheta = \lambda$ and $h(x) = f(x, u(x))$ in (5.7). We obtain immediately $u(x) \geq 0$ for a.e. $x \in \Omega$.

∎

Example 6.1 Let us consider the BVP

$$-\operatorname{div}(w(x)e^{u^2}|\nabla u|^2\nabla u) = \frac{1}{c\pi^2}u^3\arctan^2(u) + \frac{1}{3c}u^3 \quad \text{in } \Omega, \tag{6.21}$$

$$u = 0 \quad \text{on } \partial\Omega.$$

where $w(x)$ is the weight function satisfying (2.6), c is the constant of the imbedding $W_0^{1,4}(w,\Omega) \hookrightarrow L^4(\Omega)$. In this case we have $p = 4$, $\lambda = \frac{1}{c\pi^2}$, $a(x,s) = w(x)e^{s^2}$, $b(x,s) = s^3\arctan^2(s)$, $f(x,s) = \frac{1}{3c}s^3$, $\alpha(x) = \frac{\pi^2}{4}$, $\beta(x) \equiv \frac{1}{3c}$, $\gamma(x) \equiv 0$. It is possible to show that the hypotheses of Theorem 6.1 are satisfied. Hence the BVP (6.21) has at least one weak solution $u \in W_0^{1,4}(w,\Omega) \cap L^\infty(\Omega)$.

□

Example 6.2 Let $\Omega = (0,\pi) \times (0,\pi)$ be the plane domain, $x = (x_1, x_2) \in \mathbb{R}^2$. Define

$$w(x) = |\frac{x_1}{\pi}|^\mu |\frac{x_2}{\pi}|^\nu, \quad \mu, \nu \in \mathbb{R}, \quad x \in \mathbb{R}^2,$$

and consider the BVP

$$-\operatorname{div}(w(x)(1+u^4)\nabla u) = (\frac{u}{2\pi} + 1)\operatorname{arccotan}(u) \quad \text{in } \Omega, \tag{6.22}$$

$$u = 0 \quad \text{on } \partial\Omega.$$

In this case we have $n = p = 2$, $\lambda = 1$, $a(x,s) = w(x)(1+s^4)$, $b(x,s) = \frac{s}{2\pi}\operatorname{arccotan}(s)$, $f(x,s) = \operatorname{arccotan}(s)$, $\alpha(x) \equiv \frac{1}{2}$, $\beta(x) \equiv 0$, $\gamma(x) \equiv \pi$. We will assume

$$\mu < 0, \quad \nu < 0. \tag{6.23}$$

Then (6.23) guarantees the validity of (2.6). Since $\lambda_1 = 1$ is the first eigenvalue of the homogeneous Dirichlet problem for the Laplace operator on $\Omega = (0,\pi) \times (0,\pi)$, the condition (6.23) implies

$$\int_0^\pi \int_0^\pi u^2 dx_1 dx_2 \leq \int_0^\pi \int_0^\pi |\nabla u|^2 dx_1 dx_2 \leq \int_0^\pi \int_0^\pi w(x_1,x_2)|\nabla u|^2 dx_1 dx_2$$

for any $u \in W_0^{1,2}(w,\Omega)$. Hence we can put $c = 1$ for the imbedding constant $W_0^{1,5}(w,\Omega) \hookrightarrow L^2(\Omega)$. Thus $\lambda^\# \geq 2$ and the assumptions of Theorem 6.2 are fulfilled. Hence the BVP (6.22) has at least one weak solution $u \in W_0^{1,2}(w,\Omega) \cap L^\infty(\Omega)$ such that $u \geq 0$ a.e. in Ω. Arguing as in Remarks 3.2 and 4.4 we have even $u > 0$ in Ω in this case.

□

References

[1] R. A. Adams: *Sobolev Spaces*, Academic Press, Inc., New York 1975.

[2] A. Anane: *Simplicité et isolation de la premiére valeur propre du p-laplacien avec poids*, C. R. Acad. Sci. Paris Sér. I Math. 305 (1987), 725-728.

[3] G. Barles: *Remarks on uniqueness results of the first eigenvalue of the p-Laplacian*, Ann. Fac. des Sc. de Toulouse IX, no 1 (1988), 65-75.

[4] T. Bhattacharya: *Radial symmetry of the first eigenfunction for the p-Laplacian in the ball*, Proc. Amer. Math Society 104 (1988), 169-174.

[5] L. Boccardo: *Positive eigenfunctions for a class of quasi-linear operators*, Bollettino U. M. I. (5) 18-B (1981), 951-959.

[6] F. E. Browder, W. V. Petryshin: *Approximation methods and the generalized topological degree for nonlinear mappings in Banach spaces*, J. Func. Analysis 3, (1969), 217-245.

[7] P. Drábek: *Solvability and Bifurcations of Nonlinear Equations*, Pitman Research Notes 232, Longman, Essex 1992.

[8] P. Drábek, M. Kučera: *Generalized eigenvalue and bifurcations of second order boundary value problems with jumping nonlinearities*, Bull. Austral. Math. Society 37 (1988), 179-187.

[9] P. Drábek, A. Kufner, F. Nicolosi: *On the solvability of degenerated quasilinear elliptic equations of higher order*, Journal of Differential Equations 109 (1994), 325-347.

[10] J. Fleckinger, J. Hernández, F. De Thélin: *Principe du maximum pour un systéme elliptique non linéaire*, C. R. Acad. Sci. Paris, t. 314, Sér. I (1992), 665-668.

[11] S. Fučík, A. Kufner: *Nonlinear differential Equations*, Elsevier, The Netherlands 1980.

[12] J. García Azorero, I. Peral Alonso: *Existence and non-uniqueness for the p-Laplacian: Non-linear eigenvalues*, Comm. Partial Differential Equations 12 (1987), 1389-1430.

[13] M. A. Krasnoselskij: *Positive solutions of operator equations* (Russian), Moscow 1962. English translation: P. Noordhoff, Groningen.

[14] M. G. Krein, M. A. Rutman: *Linear operators leaving invariant a cone in a Banach space*, Amer. Math. Soc. Translations, Ser. 1, 10 (1950), 199-325.

[15] A. Kufner, O. John, S. Fučík: *Function Spaces*, Academia, Prague 1977.

[16] A. Kufner, A. M. Sändig: *Some Applications of Weighted Sobolev Spaces*, Teubner, Band 100, Leipzig 1987.

[17] P. Linqvist: *On the equation* $\operatorname{div}(|\nabla u|^{p-2}\nabla u) + \lambda |u|^{p-2}u = 0$, Proc. Amer. Math. Society 109 (1990), 157–164.

[18] M. Otani, T. Teshima: *The first eigenvalue of some quasilinear elliptic equations*, Proc. Japan Academy 64, sér. A (1988), 8–10.

[19] I. V. Skrypnik: *Nonlinear Elliptic Boundary Value Problems* (Russian), Naukovaja Dumka, Kyjev 1973 (English translation: Teubner, Leipzig 1986).

[20] P. Tolksdorf: *Regularity for a more general class of quasilinear elliptic equations*, J. Differential Equations 51 (1984), 126–150.

[21] N. S. Trudinger: *On Harnack type inequalities and their application to quasilinear elliptic equations*, Comm. Pure Appl. Math. 20 (1967), 721–747.

[22] M. M. Vainberg: *Variational Methods for the Study of Nonlinear Operators*, Holden–Day, Inc. San Francisco 1964.

Department of Mathematics
University of West Bohemia
P.O.BOX 314, 306 14 Plzeň
Czech Republic
e-mail: pdrabek@kma.zcu.cz

M GIRARDI AND M MATZEU

Some results about periodic solutions of second order Hamiltonian systems where the potential has indefinite sign

The aim of this paper is to present some results obtained by the authors (see [5], [6], [7]) about periodic solutions of second order nonautonomous Hamiltonian systems where the potential has indefinite sign and has a superquadratic growth in the state variable.

More precisely, we shall consider the following system, for a fixed $T > 0$,

(V) $\quad \begin{cases} \ddot{x}(t) + b(t)V'(x(t)) = 0 \,, \ x(t) \in \mathbf{R}^N \\ x(0) = x(T), \dot{x}(0) = \dot{x}(T) \end{cases}$

where

(b_1) $b(\cdot)$ is a T-periodic continuous real function such that
$\exists \bar{t} \in [0, T] : b(\bar{t}) > 0$

(V_1) $\qquad V \in C^2(\mathbf{R}^N), V(x) \geq 0 \forall x \in \mathbf{R}^N$

(V_2) $\qquad V(x) = o(|x|^2)$ at $x = 0$

(V_3) $\qquad \exists \beta > , \ R > 0 : \beta V(x) \leq V'(x)x \qquad \forall x \in \mathbf{R}^N, |x| \geq R \,.$

In case that $b(t) > 0 \ \forall t \in [0, T]$, it s well known that there exists a nontrivial T-periodic solution of (V) (see e.g. [9] Robinowitz).

A first existence result in case that $b(\cdot)$ is negative somewhere was due to Lassoued in [8], where V is supposed to be homogeneous and striatly convex. In the following, Ben Naoum, Troestler and Willem obtained in [2] the same result without the convexity assumption but always in the homogeneous case (indeed in [2] a subquadratic case too is considered).

Howewer let us point that the techniques used in these papers require the necessity of the homogeneity condition. Actually this is an appropriate property which guarantees that the functional suitably associated with (V) satisfies the Palais-Smale condition.

A first possibility of weakening the homogeneity assumption in such a way that the Palais-Smale still holds is to require that V differs from a homogeneous superquadratic function only by a quadratic term outside of a sufficiently large sphere. This requirement is connected with a control put on the negative part $b^-(\cdot)$ of b in dependence of the superquadratic growth of V. By this assumption one is able to generalize just the classical result related to the case $b^- = 0$.

Also one can state a multiplicity result if V is even and the other assumptions on V are given on the whole space.

More precisely one gets the following existence results

Theorem 1. *(see [5], [6]). Let $b(\cdot)$ satisfy (b_1), and*

(b_2) $\int_0^T b(t) > 0$

and let V satisfy (V_1), (V_2), (V_3). If there exist two numbers $c \geq 0$, $d > 0$ such that

(1) $B^-(V'(x)x - \beta V(x)) \leq c|x|^2$ *for* $|x| \geq R$, *with* $B^- = \max\{b^-(t) : t \in \mathbf{R}\}$

(2) $c < \frac{2(\beta-2)}{(1+4\pi^2)} \frac{\pi^2}{T^2}$

(3) $B^-(|V''(x)| - d|x|^{\beta-2}) \leq 0$ *for* $|x| \geq R$,

then there exists a non-zero T-periodic solution of (V).

Theorem 2. *(see [5], [6]). Let $b(\cdot)$ satisfy (b_1) and*

(b_3) $\int_0^T b(t) < 0$

and let V satisfy (V_1), (V_2), (V_3) and moreover

(V_4) $\exists a_1 > 0$, $\beta' > 2 : V(x) \geq a_1|x|^{\beta'}$ *if* $|x| \leq r$

(V_5) $\exists a_2 > 0 : |V'(x)| \leq a_2|x|^{\beta'-1}$ *if* $|x| \leq r$.

If there exist two numbers $c \geq 0$, $d > 0$ such that (1), (2), (3) hold, then there exists a non-zero T-periodic solution of (V).

The proofs of these two theorems are based on the well known fact that the T-periodic solutions of (V) can be obtained by periodically extending to the whole real line the critical points of the functional

$$F(u) = \frac{1}{2}\int_0^T |\dot{u}|^2 - \int_0^T b(t)V(u)$$

on the space

$$H_T^1 = \{v \in H^1(0, T; \mathbf{R}^N) : v(0) = v(T)\}.$$

At this purpose a basic lemma in the proof of theorems 1.2 is given by the following result, which seems to be interesting by itself.

Lemma 1. *Let $b(\cdot)$ satisfy (b_1) and let*

$$\int_0^T b(t) \neq 0 .$$

Let V satisfy (V_1), (V_3). If, moreover, there exist two numbers $c \geq 0$, $d > 0$ such that (1), (2), (3) hold, then f satisfy the Palais-Smale condition on H_T^1.

At this point, in case that (b_2) holds, one checks that the functional f is positive on a sufficiently small sphere of the subspace $\tilde{H}_T^1$ of H_T^1 given by

$$\tilde{H}_T^1 = \left\{ u \in H_T^1 : \int_0^T u = 0 \right\}$$

and that f is negative on the constant functions. Starting from these remarks one is able to discover a geometrical structure of linking type for the functional f around the origin of H_T^1, so Lemma 1 enables to exhibit a critical point of linking type.

On the other side, when (b_3) holds, it is easy to see that f is positive on the constant functions, as well as on a suitable small sphere of $\tilde{H}_T^1$ and f is negative somewhere in H_T^1.

Then a careful analysis of the different behaviours of f on the constant functions and on $\tilde{H}_T^1$ and still Lemma 1 allow to get a critical point of Mountain Pass type.

The mentioned multiplicity result is given by the following

Theorem 3. *(see [5], [6]). Let $b(\cdot)$ satisfy (b_1) and*

(b_4) $b(t) = b(T-t)$ $\forall t \in [0, T/2]$,

and let $V \in C^2(\mathbf{R}^N)$ satisfy

(V_6) $V(-x) = V(x)$ $\forall x \in \mathbf{R}^N$

(V_7) $\exists a_1 > 0$, $\beta > 2 : V(x) \geq a_1 |x|^\beta$ $\forall x \in \mathbf{R}^N$

(V_8) $\exists a_2 > 0 : |V'(x)| \leq a_2 |x|^{\beta-1}$ $\forall x \in \mathbf{R}^N$

(V_9) $\beta V(x) \leq V'(x)x$ $\forall x \in \mathbf{R}^N$

(V_{10}) $V''(x)xx \geq (\beta - 1)V'(x)x$ $\forall x \in \mathbf{R}^N$.

If there exist two numbers $c' \geq 0$, $d > 0$ such that

(4) $B^- (V'(x)x - \beta V(x)) \leq c' |x|^2$ $\forall x \in \mathbf{R}^N$

(5) $B^- (V''(x)xx - (\beta - 1)V'(x)x) \leq d' |x|^2$ $\forall x \in \mathbf{R}^N$

(6) $\max(2c', d') < \frac{8(\beta-2)}{T^2}$

then there exist infinitely many pairs[(*)] of distinct T-periodic solutions x of (V) satisfying

(7)
$$x\left(t + \frac{T}{2}\right) = -x\left(\frac{T}{2} - t\right) \quad \forall t \in \left[0, \frac{T}{2}\right].$$

The proof of Theorem 3 is based on the consideration of the Nehari's manifold M associated with the functional

$$f(u) = \frac{1}{2} \int_0^{T/2} |\dot{u}|^2 - \int_0^{T/2} b(t) V(u)$$

on the space $X = H_0^1(0, T/2; \mathbf{R}^N)$, whose critical points yield in a standard way, T-periodic solutions of (V) satisfying (7). The manifold M is defined as

$$M = \{u \in X \smallsetminus \{0\} :< f'(u), u >= 0\}$$

(where $< \cdot, \cdot >$ denotes the duality pairing between X and its dual space).

The assumptions of Theorem 3 enable to state that M is a closed regular manifold of X and that the critical points of f on X just coincide with the critical points of $f_{|M}$, the restriction of f on M.

Moreover, some minimization procedures and the application of a basic result of the Lusternik-Schrinelman category theory enable to find infinitely many critical points of f on M, so infinitely many pairs of solutions of (V), satisfying (7).

Actually, condition (1) seems to be a little restrictive if one looks at some very general cases. From one side, it implies a bound for the negative part of $b(\cdot)$, on the other side not even the simple sum of two homogeneous superquadratic functions having different homogeneity degrees could be choosen as the potential V, since it does not verify condition (1).

Indeed the real difficulty, if one uses the variational approach, is the check of the Palais-Smale condition.

A possible way of bypassing this problem is to use a suitable truncature argument in such a way that the "truncated" potential is homogeneous at infinity (so the P.S.-condition is satisfied) and then to apply some a priori estimated deduced by the consideration of the Morse index related to the solutions of the "truncated" problems.

This procedure and a suitable finite-dimensional approximation of the problem have been used in a recent paper [7] and carry to the following result

---

[(*)]Note that, due to (V_5), if x is a solution of (V), then $-x$ too is so, then it is right to speak of "pairs" of solutions.

Theorem 4. (see [7]). *Let $b(\cdot)$ satisfy (b_1), (b_3) and (b_5) the set*

$$\mathcal{Z} = \{t \in [0,T] : b(t) = 0 \text{ and } \exists \varepsilon = \varepsilon(t) \text{ s.t.} \\ b(\cdot) > 0 \text{ either in } (t-\varepsilon, t) \text{ or in } (t, t+\varepsilon)\}$$

is not empty and finite
(b_6) *the function $b^+(t) = \max(b(t), 0)$ belongs to $H^1([0,T], \mathbf{R}^N)$.*
Let $V = V_1 + \cdots + V_{m-1} + V_m$ $(m \geq 2)$ such that
(V_i) *V_i is positive outside of the origin of $\mathbf{R}^N$ and positively homogeneous with degree $\beta_i > 3$ $(i = 1, \ldots, m)$*
(V_m) *$\exists R_1 > 0$ such that, putting $\beta_m = \max\{\beta_i : i = 1, \ldots, m\}$, V_m is strictly convex for $|x| \geq R_1$.*
Then, if $\beta_i < 6$ for $i = 1, \ldots, m-1$, and $\beta_m > 4$, there exists at least one non-zero T-periodic solution of (V).

Let us make two remarks about the hypotheses of this result. Firstly one notes that no upper bound is imposed on the negative part $b^-(\cdot)$ of $b(\cdot)$ ($b^-(\cdot) = b(\cdot) - b^+(\cdot)$), contrarily to the case of theorems 1, 2. Secondly let us point out that a large class of polynomia in the $|x|$-variable statisfies the assumptions of the potential V. More precisely V can be choosen as $V(x) = P(|x|) \; \forall x \in \mathbf{R}^N$, where

$$P(s) = a_4 s^4 + a_5 s^5 + a_m s^m \qquad s \in \mathbf{R}$$

with an arbitrary integer $m \geq 6$ and arbitrary numbers $a_m > 0$, $a_4 \geq 0$, $a_5 \geq 0$ with $a_4 a_5 > 0$.

Let us give now a short sketch of the proof of Theorem 4. As a first step, putting

$$\overline{V}(x) = V_1(x) + \cdots + V_{m-1}(x) \qquad \forall x \in \mathbf{R}^N ,$$

one considers, for arbitrarily fixed $R > 0$, a "truncated" potential V_R of the form

$$V_R(x) = \overline{V}(x) + \chi_R(|x|) V_m(x)$$

where χ_R is a $C^2(\mathbf{R}_+)$ function such that
(8)
$$\chi_R(t) = 1 \; \forall t \in [0,R], \quad \chi_R(t) = 0 \; \forall t \in [2R, +\infty), \quad 0 \leq \chi_R(t) \leq 1 \; \forall t \in \mathbf{R}_+$$

(9)
$$|\chi'_R(t)| \leq \lambda_1/R \quad \forall t \in \mathbf{R}_+ , \quad \text{for some } \lambda_1 > 0$$

(10)
$$|\chi''_R(t)| \leq \lambda_2/R^2 \quad \forall t \in \mathbf{R}_+ , \quad \text{for some } \lambda_2 > 0$$

and consider the associated functional

$$f_R(v) = \frac{1}{2}\int_0^T |\dot{v}|^2 - \int_0^T b(t) V_R(v)$$

on a suitable sequence $\{E_n\}_{n\in\mathbb{N}}$ of finite dimensional subspaces of the space $H_T^1 = \{v \in H^1(0,T;\mathbb{R}^N) : v(0) = v(T)\}$ such that $H_T^1 = \overline{\bigcup_{n\in\mathbb{N}} E_n}$.

Then it is easy to show that f_R satisfies the Palais-Smale condition on H_T^1, so on each E_n too. Therefore, as (b_3) holds, one can use the same arguments as in the proof of Theorem 2 in order to find a critical point of Mountain Pass type u_R^n for the restriction f_R^n of f_R on E_n such that

(11) $\qquad i_R^n(u_R^n) \leq 1 \qquad$ where $i_R^n(u_R^n)$ denotes the Morse index of u_R^n as a critical point of f_R^n

Moreover, due to the particular structure of E_n and the properties of χ_R, one gets

(12) $\qquad f_R^n(u_R^n) \leq c_1 \qquad$ a constant independent of R, n

At this point (b_2), (b_6), (11), the growth properties of V_R and condition (V_m) enable to prove the estimate

(13) $\qquad \|u_R^n\|_{L^2(\text{supp } b+)} \leq c_2 \qquad$ a constant independent of R, n ,

where supp b^+ denotes the support of the function $b^+(\cdot)$ in the interval $[0,T]$.

Secondly (12), (13) and the equivalence of all the norms in the finite-dimensional space E_n yield, for any $n \in \mathbb{N}$,

(14) $\qquad \|u_R^n\|_{H_T^1} \leq d_n$, a constant independent of R ,

which allows to state that, for sufficiently large R, $u^n = u_R^n$ is critical for the restriction of the "original" functional f, that is

$$f(v) = \frac{1}{2}\int_0^T |\dot{v}|^2 - \int_0^T b(t) V(v)$$

on the space E_n.

As a third step, an estimate of the type

(15) $\qquad \|u^n\|_{H_T^1} \leq c_3$, a constant independent of n ,

can be obtained as a consequence of (13) and a suitable use of the Gagliardo-Niremberg inequality (see [3] e.g.) based on the property $\beta_i < 6$ for $i = 1,\ldots,m-1$. Obviously (15) enables to find, by passing to the limit of a subsequence of $\{u^n\}$, a solution u of (15). Finally, the non-triviality of u is implied by the estimate from below

$$\|u^n\|_{H^1_T} \geq c_4 > 0$$

which derives from the Mountain Pass nature of u^n and the relation

$$f(v) \geq c_5 > 0 \quad \forall v \in H^1_T : \|v\| = \rho, \quad \text{with } \rho > 0 \text{ sufficiently small}$$

deduced from the growth behaviour of V at the origin.

Finally let us mention a result about subharmonic solutions of (V) in case that V is homogeneous.

Theorem 5. *Let $b(\cdot)$ satisfy (b_1), (b_4) and*

(b_7) *the set*

$$Z' = \{t \in [0, T/2] : b(t) = 0 \text{ and } \exists \delta = \delta(t) \text{ s.t. } b(t_1)b(t_2) < 0$$
$$\forall t_1 \in (t - \delta, t) \quad \forall t_2 \in (t, t + \delta)\}$$

is finite.

Let $V \in C^2(\mathbf{R}^N)$ be positively homogeneous with degree $\beta > 2$ and $V(x) > 0 \ \forall x \in \mathbf{R}^N \setminus \{0\}$, and let (V_6) hold. Then, for any $k \in \mathbf{N}$, there exist infinitely many pairs of distinct kT-periodic solutions $x^{(k)}$ of (V) such that

$$x^{(k)}(t + kT/2) = -x^{(k)}(kT/2 - t) \ \forall t \in [0, kT/2]$$

Moreover, for any odd integer k, at least two distinct pairs of these solutions have minimal period kT.

The proof is based on the consideration of the same variational framework illustrated in the proof of Theorem 3, only by replacing the space $X = H^1_0(0, T/2; \mathbf{R}^N)$ with the space $X_k = H^1_0(0, kT/2; \mathbf{R}^N)$ and the manifold M with the manifold M_k defined in the same way as M but in the space X_k. In the present case, one is able to prove, by some delicate arguments with a lot of technicalities, that M_k is path connected, due to the homogeneity of V, so one can construct a pair of minimum points of $f_{|M_k}$ (say $u_k, -u_k$) as well as a critical point of Mountain Pass type for $f_{|M_k}$, say $\tilde{u}_k$. It is easy to check, by straightforward arguments, the minimality of the period kT for the solution $x^{(k)}$ corresponding to the minimum point of u_k, while the same property for the solution $\tilde{x}^{(k)}$ corresponding to $\tilde{u}_k$ relies on an appropriate use of the usual Morse index estimates for critical points of Mountain Pass type due to Ekeland and Hofer (see e.g. [4]), suitable adapted to this "constrained" case.

REFERENCES

[1] A. Ambrosetti, P. Rabinowitz, "Dual variational methods in critical point theory and applications", *J. Funct. Anal.*, **14** (1973), 349-381.

[2] A.K. Ben Naoum, C. Troestler, M. Willem, "Existence and multiplicity results for homogeneous second order differential equations", to appear on *J. Diff. Eq.*

[3] M. Brezis, *Analyse fonctionnelle: théorie et applications*, Masson (1983).

[4] I. Ekeland, *Convexity methods in Hamiltonian mechanics*, Springer-Verlag (1990).

[5] M. Girardi, M. Matzeu, "Existence and multiplicity results for periodic solutions of superquadratic Hamiltonian systems where the potential changes sign", to appear on *Nonlinear Diff. Eq. and Appl.*

[6] M. Girardi, M. Matzeu, "Periodic solutions of second order nonautonomous systems with the potential changing sign", *Rend. Mat. Acc. Lincei*, s. 9, v. 4 (1993), 273-277.

[7] M. Girardi, M. Matzeu, "On periodic solutions of a class of second order nonautonomous systems with nonhomogeneous potentials indefinite in sign", preprint.

[8] L. Lassoued, "Periodic solutions of a second superquadratic system with change of sign of the potential", *J. Diff. Eq.*, **93** (1991), 1-18.

[9] P. Rabinowitz, "Periodic solutions of Hamiltonian systems", *Comm. Pure Appl. Math.* **31** (1978), 157-184.

Address

M. Girardi, Dipartimento di Matematica dell'Università di Roma III, Via C. Segre 2/6, 00146 Roma (Italy).

M. Matzeu, Dipartimento di Matematica dell'Università di Roma Tor Vergata, Viale della Ricerca Scientifica, 00133 Roma (Italy).

J P GOSSEZ AND M MOUSSAOUI
A note on nonresonance between consecutive eigenvalues for a semilinear elliptic problem

Introduction

This paper is concerned with the semilinear elliptic problem

(1.1) $$\begin{cases} -\Delta u = \lambda_k u + f(x,u) + h(x) & \text{in } \Omega, \\ u = 0 & \text{on } \partial\Omega, \end{cases}$$

where Ω is a bounded open subset of $\mathbf{R}^N$, $f : \Omega \times \mathbf{R} \to \mathbf{R}$ is a Carathéodory function, h is a function on Ω, and λ_k, $k = 1, 2, \ldots$, denote the (ordered distinct) eigenvalues of $(-\Delta)$ on $H_0^1(\Omega)$. We are interested in the conditions to be imposed on the nonlinearity f in order to guarantee nonresonnance, i.e, the solvavility of (1.1) for "any h".

Beginning with Dolph [4], several works have been concerned with the study of this question. Various conditions on the asymptotic behaviour of the quotients $\frac{f(x,s)}{s}$ and $\frac{2F(x,s)}{s^2}$ have been introduced. Here $F(x,s)$ denotes the primitive $\int_0^s f(x,t)dt$, and we will write

$$k_\pm(x) = \liminf_{s \to \pm\infty} \frac{f(x,s)}{s}, \quad \ell_\pm(x) = \limsup_{s \to \pm\infty} \frac{f(x,s)}{s},$$

$$K_\pm(x) = \liminf_{s \to \pm\infty} \frac{2F(x,s)}{s^2}, \quad L_\pm(x) = \limsup_{s \to \pm\infty} \frac{2F(x,s)}{s^2},$$

with, for an autonomous nonlinearity $f(x,s) = f(s)$, $k_\pm$ instead of $k_\pm(x)$. Dolph showed in [4] the solvability of (1.1) for any $h \in L^2(\Omega)$ if, for some constants μ_k and μ_{k+1},

(1.2) $$0 < \mu_k \leq k_\pm(x) \leq \ell_\pm(x) \leq \mu_{k+1} < \lambda_{k+1} - \lambda_k.$$

He also derived the same conclusion under the weaker condition

(1.3) $$0 < \mu_k \leq K_\pm(x) \leq L_\pm(x) \leq \mu_{k+1} < \lambda_{k+1} - \lambda_k$$

with however, in this latter case, an additional rather technical assumption bearing on the functional associated to (1.1). The situation where $k_\pm = 0$ and $\ell_\pm = \lambda_{k+1} - \lambda_k$ was considered (in the autonomous case) in [3], where a so called positive density condition was introduced. Roughly speaking, this condition imposes to $\frac{f(s)}{s}$ to remain > 0 and $< \lambda_{k+1} - \lambda_k$ for sufficiently many values of s as $s \to \pm\infty$. In [8], it was later proved that this positive density condition is in fact equivalent to $0 < K_\pm$ and $L_\pm < \lambda_{k+1} - \lambda_k$, which shows the connection with (1.3). More recently Costa and Oliveira [2] extended the result of [3] to the non autonomous case, allowing equality in both sides of (1.2) for a.e $x \in \Omega$, and also in both sides of (1.3) with however there strict inequality on a subset of positive measure.

In this paper we are interested in situations where both $k_\pm(x)$ and $K_\pm(x) = 0$ for a.e $x \in \Omega$ (or both $\ell_\pm(x) = \lambda_{k+1} - \lambda_k$ and $L_\pm(x) = \lambda_{k+1} - \lambda_k$ for a.e $x \in \Omega$). It is clear that in such situations the solvability of (1.1) cannot be guaranteed without further assumption on the nonlinearity f. We shall impose some growth condition on f together with some variant of the Ahmad-Lazer-Paul condition on the primitive F.

To state our main result, let us denote by $E(\lambda_i)$ the λ_i-eigenspace, by $\|.\|_1$ the $H_0^1(\Omega)$ norm, and by $\|.\|_0$ the $L^2(\Omega)$-norm.

Theorem 1.1. *Assume:*

(f_0) $\qquad \forall R > 0 \quad \sup_{|s| \leq R} |f(x,s)| \in L^2(\Omega);$

(f_1) $\qquad \limsup_{s \to \pm\infty} \dfrac{f(x,s)}{s} \leq \ell(x) \leq \lambda_{k+1} - \lambda_k$ *uniformly in x with*

$$\text{meas }\{x \in \Omega / \ell(x) < \lambda_{k+1} - \lambda_k\} > 0;$$

(f_2) $\qquad f(x,s) \geq -as^\alpha - b(x), x \in \Omega, s \geq 0,$

$$f(x,s) \leq a|s|^\alpha + b(x), x \in \Omega, s \leq 0,$$

where $a \geq 0, b(x) \in L^2(\Omega)$ and $0 \leq \alpha < 1$;

(f_3) $\qquad \dfrac{1}{\|u^0\|_1^{1+\alpha}} \displaystyle\int_\Omega F(x, u^0(x))dx \to +\infty$ *as* $\|u^0\|_1 \to +\infty, u^0 \in E(\lambda_k).$

Then, for any $h \in L^2(\Omega)$, problem (1.1) has at least one solution.

Several works have been devoted recently to the study of (1.1) when both $k_\pm(x) = 0$ and $K_\pm(x) = 0$ a.e. In particular Theorem 1.1 extends the result of Ramos [12] which corresponds to the case $\alpha = 0$ and $\ell(x) \leq \mu < Min\left(\dfrac{\lambda_k}{\lambda_{k-1}} - 1, \lambda_{k+1} - \lambda_k\right)$. Costa [1] and Silva [13] also obtained an existence result for problem (1.1) under the hypothesis

(1.4) $\qquad |f(x,s)| \leq a|s|^\alpha + b(x) \qquad x \in \Omega, s \in \mathbb{R},$

where $b(x) \in L^2(\Omega)$ and $0 \leq \alpha < 1$, and

(1.5) $\qquad \dfrac{1}{\|u^0\|_1^{2\alpha}} \displaystyle\int_\Omega F(x, u^0(x))dx \to \pm\infty$ as $\|u^0\|_1 \to +\infty, u^0 \in E(\lambda_k).$

Theorem 1.1 can be looked at as a partial extension of this result since (1.4) clearly implies (f_1)-(f_2). Other results less directly connected with Theorem 1.1 but also concerned with situations where both $k_\pm(x) = 0$ and $K_\pm(x) = 0$ a.e can be found in [6], [7], [10].

Theorem 1.1 is proved in section 2. The proof is variational and based on the saddle point theorem of Rabinowitz. Some corollaries and variants are presented in section 3. In particular we consider there some more explicit conditions on F which guarantee the validity of (f_3). In section 4 we indicate several examples where our results apply and where, as far as we can see, previously known results do not.

Proof of Theorem 1.1.

The functional associated to (1.1) is

$$\Phi(u) = \frac{1}{2}\|u\|_1^2 - \frac{\lambda_k}{2}\|u\|_0^2 - \int_\Omega F(x,u)dx - \int_\Omega hu\,dx.$$

Under conditions (f_0), (f_1) and (f_2), it is well know that Φ is well defined on $H_0^1(\Omega)$, weakly lower semicontinuous and C^1. Its critical points are the weak solutions of (1.1).

Write $H_0^1(\Omega) = E^- \oplus E^0 \oplus E^+$, with $E^- = \oplus_{i \leq k-1} E(\lambda_i)$, $E^0 = E(\lambda_i)$, and $E^+ = (E^- \oplus E^0)^\perp$, where the orthogonality is taken with respect to the scalar product associated to $-\Delta$ on $H_0^1(\Omega)$:

$$(u,v) = \int_\Omega \nabla u \nabla v \, dx.$$

We recall that

(2.1) $$\int_\Omega |\nabla u^-|^2 dx \leq \lambda_{k-1} \int_\Omega (u^-)^2 dx \quad , \forall u^- \in E^-,$$

(2.2) $$\int_\Omega |\nabla u^+|^2 dx \geq \lambda_{k+1} \int_\Omega (u^+)^2 dx \quad , \forall u^+ \in E^+.$$

We begin with the following two lemmas.

Lemma 2.1. *Under conditions (f_0), (f_1), there exists $\delta > 0$ such that*

$$\|u^+\|_1^2 - \lambda_k\|u^+\|_0^2 - \int_\Omega \ell(x)(u^+)^2 > 2\delta\|u\|_1^2$$

for all $u^+ \in E^+$.

Proof: It is similar to the proof of lemma 1 in [9]. Q.E.D.

Lemma 2.2. *If f satisfies (f_0), (f_1) and (f_2), then there exist three measurable functions $\beta, \gamma : \Omega \times \mathbf{R} \to \mathbf{R}$, $c : \Omega \to \mathbf{R}$ such that*
(i) $0 \leq \beta(x,s) \leq (\ell(x) + \delta\frac{\lambda_{k+1}}{2})$,
(ii) $c(x) \in L^2(\Omega)$ and $|\gamma(x,s)| \leq a|s|^\alpha + c(x)$
(iii) $f(x,s) = \beta(x,s)s + \gamma(x,s)$
for every $s \in \mathbf{R}$, $x \in \Omega$, where δ is given by lemma 2.1.

Proof: From (f_0) and (f_1), there exists $d(x) \in L^2(\Omega)$ such that

(2.3) $$f(x,s) \leq \left(\ell(x) + \frac{\delta\lambda_{k+1}}{2}\right)s + d(x) \text{ for } x \in \Omega, s \geq 0,$$

(2.4) $$f(x,s) \geq \left(\ell(x) + \frac{\delta\lambda_{k+1}}{2}\right)s - d(x) \text{ for } x \in \Omega, s \leq 0.$$

Define
$$\beta(x,s) = \begin{cases} Max\left(\frac{f(x,s)-d(x)}{s};0\right) & s > 0 \\ 0 & s = 0 \\ Max\left(\frac{f(x,s)+d(x)}{s};0\right) & s < 0 \end{cases}$$

and $\gamma(x,s) = f(x,s) - \beta(x,s)s$. Chosing $c(x) = |b(x)| + |d(x)|$, it is easy to see that the three properties stated in the lemma are satisfied. Q.E.D.

We will now study the shape of the functional Φ.

Lemma 2.3. *The restrictions $(-\Phi)|_{E^-\oplus E^0}$ and $\Phi|_{E^+}$ are coercive, that is,*
(i) $\Phi(u) \to -\infty$ as $\|u\|_1 \to +\infty$, $u \in E^- \oplus E^0$,
(ii) $\Phi(u) \to +\infty$ as $\|u\|_1 \to +\infty$, $u \in E^+$.

Proof: (i) When $k = 1$, the conclusion follows directly from (f_3). So let us assume $k \geq 2$.

For $u \in E^- \oplus E^0$, write $u = u^- + u^0$, with $u^- \in E^-$ and $u^0 \in E^0$. We will use the following inequality, which is readily verified:

(2.5) $$\left(\frac{a}{2^{m+1}} - b\right)^2 + \left(\frac{a}{2^{m+1}} - b\right)b \leq \frac{1}{2^m}(b-a)^2 \quad \forall a, b \in \mathbf{R}, \forall m \in \mathbf{N}$$

(the particar case $m = 0$ was already considered in [12]). Let us fix m such that

(2.6) $$\frac{1}{2^m}\left(\lambda_{k+1} - \lambda_k + \frac{\delta\lambda_{k+1}}{2}\right) \leq \frac{1}{4}(\lambda_k - \lambda_{k-1}).$$

Write

$$\Phi(u) = q(u^-) - \int_\Omega F\left(x, \frac{u^0}{2^{m+1}}\right) dx + \int_\Omega \left[F\left(x, \frac{u^0}{2^{m+1}}\right) - F(x,u)\right] dx + \int_\Omega hu\, dx$$

where $q(u) = \frac{1}{2}\|u\|_1^2 - \frac{\lambda_k}{2}\|u\|_0^2$. From Lemma 2.2 and (2.5), we have

$$F\left(x, \frac{u^0}{2^{m+1}}\right) - F(x,u) = \left(\frac{u^0}{2^{m+1}} - u\right)\int_0^1 f\left(x, u + t\left(\frac{u^0}{2^{m+1}} - u\right)\right) dt$$

$$= \left(\frac{u^0}{2^{m+1}} - u\right)\int_0^1 A(t)\left(u + t\left(\frac{u^0}{2^{m+1}} - u\right)\right) dt +$$

$$\left(\frac{u^0}{2^{m+1}} - u\right)\int_0^1 \gamma\left(x, u + t\left(\frac{u^0}{2^{m+1}} - u\right)\right) dt$$

$$= \left(\frac{u^0}{2^{m+1}} - u\right)^2 \int_0^1 tA(t)dt + \left(\frac{u^0}{2^{m+1}} - u\right) u \int_0^1 A(t)dt +$$

$$\left(\frac{u^0}{2^{m+1}} - u\right) \int_0^1 \gamma\left(x, u + t\left(\frac{u^0}{2^{m+1}} - u\right)\right) dt$$

$$\leq \left[\left(\frac{u^0}{2^{m+1}} - u\right)^2 + \left(\frac{u^0}{2^{m+1}} - u\right) u\right] \int_0^1 A(t) dt +$$

$$a(|u^0| + |u^-|) \int_0^1 \left|u + t\left(\frac{u^0}{2^{m+1}} - u\right)\right|^\alpha dt + |c(x)|(|u^0| + |u^-|)$$

$$\leq \frac{1}{2^m}(u^-)^2 \left(\lambda_{k+1} - \lambda_k + \frac{\delta \lambda_{k+1}}{2}\right) +$$

$$2a(|u^0| + |u^-|)(|u^-|^\alpha + |u^0|^\alpha) + c(x)(|u^0| + |u^-|)$$

$$\leq \frac{1}{4}(\lambda_k - \lambda_{k-1})(u^-)^2 +$$

$$2a(|u^0|^{1+\alpha} + |u^0||u^-|^\alpha + |u^0|^\alpha|u^-| + |u^-|^{1+\alpha}) +$$

$$|c(x)|(|u^0| + |u^-|)$$

where we have used that $A(t) = \beta\left(x, u + t^0\left(\frac{u^0}{2^{m+1}} - u\right)\right)$ satisfies $0 \leq A(t) \leq (\lambda_{k+1} - \lambda_k + \frac{\delta}{2}\lambda_{k+1})$. Hence it follows from Hölder and Young inequalities that

$$\int_\Omega \left[F\left(x, \frac{u^0}{2^{m+1}}\right) - F(x, u)\right] dx + \int_\Omega hu dx \leq \frac{1}{4}(\lambda_k - \lambda_{k-1})\|u^-\|_0^2 +$$
$$C_1(\|u^0\|_0^{1+\alpha} + \|u^0\|_0\|u^-\|_0^\alpha + \|u^-\|_0\|u^-\|_0^\alpha\|u^-\|_0^{1+\alpha}) + C_2(\|u^0\|_0 + \|u^-\|_0)$$
$$\leq \frac{3}{8}(\lambda_k - \lambda_{k-1})\|u^-\|_0^2 + C_3(\|u^0\|_0^{1+\alpha} + 1)$$

for some constants C_1, C_2 and C_3. Therefore

$$\Phi(u) \leq \frac{1}{2}(\lambda_{k-1} - \lambda_k)\|u^-\|_0^2 + \frac{3}{8}(\lambda_k - \lambda_{k-1})\|u^-\|_0^2 + C_3(\|u^0\|_0^{1+\alpha} + 1) -$$
$$\int_\Omega F\left(x, \frac{u^0}{2^{m+1}}\right) dx$$
$$\leq -\frac{1}{2}(\lambda_k - \lambda_{k-1})\|u^-\|_0^2 + \|u^0\|_0^{1+\alpha}\left(C_3 - \frac{1}{\|u^0\|_0^{1+\alpha}} \int_\Omega F\left(x, \frac{u^0}{2^{m+1}}\right) dx\right) + C_3.$$

Thus, by (f_3), $\Phi(u) \to -\infty$ as $\|u\|_1 \to -\infty$.

(ii) It follows from (2.3)-(2.4) that

$$F(x, s) \leq \left(\ell(x) + \frac{\delta \lambda_{k+1}}{2}\right) \frac{s^2}{2} + d(x)|s| \quad , x \in \Omega, s \in \mathbb{R},$$

hence

(2.7) $$F(x,s) \leq (\ell(x) + \delta\lambda_{k+1})\frac{s^2}{2} + d'(x) \quad, x \in \Omega, s \in \mathbb{R},$$

where $d'(x) \in L^1(\Omega)$. Using (2.7), lemmas 2.1 and 2.3, we obtain for $u \in E^+$,

$$\begin{aligned}\Phi(u) &\geq \frac{1}{2}\|u\|_1^2 - \frac{\lambda_k}{2}\|u\|_0^2 - \frac{1}{2}\int_\Omega \ell u\, dx - \frac{\delta\lambda_{k+1}}{2}\int_\Omega u^2 dx - \|d'\|_{L^1} - \|h\|_0\|u\|_0 \\ &\geq \|u\|_1^2 - \frac{\delta}{2}\|u\|_1^2 - C_1\|u\|_1 - C_2 \\ &\geq \frac{\delta}{2}\|u\|_1^2 - C_1\|u\|_1 - C_2.\end{aligned}$$

Thus Φ is coercive on E^+. Q.E.D.

Lemma 2.4. *The functional Φ satisfies the Palais-Smale condition (PS), that is, whenever $(u_n) \subset H_0^1(\Omega)$ is a sequence such that $\Phi(u_n)$ is bounded and $\Phi'(u_n) \to 0$ then (u_n) possesses a convergent subsequence.*

Proof: Assume for the moment $k \geq 2$.

Let (u_n) be a sequence as above. Since $\nabla\Phi$ a compact perturbation of the identity, it suffices to show that (u_n) is bounded (see [11], Appendice B).

Suppose, on the contrary, that for a subsequence (still denoted by u_n), $\|u_n\|_1 \to +\infty$. We decompose $u_n = u_n^- + u_n^0 + u_n^+$, where $u_n^\pm \in E^\pm$, $u_n^0 \in E^0$. To simplify notation we will omit for a moment the subscript n and write u for u_n. Proceeding as in lemma 2.3(i) and taking $m = 0$ in (2.5), we obtain

$$\begin{aligned}\Phi(u) &= q(u^-) + q(u^+) - \int_\Omega F\left(x, \frac{u^0}{2}\right) dx + \int_\Omega \left[F\left(x, \frac{u^0}{2}\right) - F(x,u)\right] dx + \int_\Omega hu\, dx \\ &\leq q(u^-) + q(u^+) + \left(\lambda_{k+1} - \lambda_k + \frac{\delta\lambda k + 1}{2}\right)\|u^+ + u^-\|_0^2 + C_1\left(\|u_0\|_0^{1+\alpha} + 1\right) - \\ & \quad \int_\Omega F\left(x, \frac{u^0}{2}\right) dx\end{aligned}$$

and so, by Poincaré inequality,

(2.8) $$\Phi(u) \leq C_2(\|u^+\|_1^2 + \|u^-\|_1^2) + C_1(\|u^0\|_1^{1+\alpha} + 1) - \int_\Omega F\left(x, \frac{u^0}{2}\right) dx$$

where the constants do not depend on u.

On the other hand, the hypothesis $\Phi'(u_n) \to 0$ yields

$$\int_\Omega \nabla u \nabla v\, dx - \lambda_k \int_\Omega uv\, dx - \int_\Omega f(x,u)v - \int_\Omega hv \leq \|v\|_1, \forall v \in H_0^1(\Omega).$$

Taking $v = u^+ - u^- - u^0$, and using lemma 2.2, we deduce

$$2q(u^+) - 2q(u^-) \leq \int_\Omega \beta(x,u)u(u^+ - u^- - u^0)dx + \int_\Omega \gamma(x,u)(u^+ - u^- - u^0)dx +$$
$$\int_\Omega h(u^+ - u^- - u^0)dx + \|u^+ - u^- - u^0\|_1$$
$$\leq \int_\Omega \beta(x,u)[(u^+)^2 - (u^- + u^0)^2]dx$$
$$+ \int_\Omega a|u^+ + u^- + u^0|^\alpha(|u^+| + |u^-| + |u^0|)$$
$$+ C_3(\|u^+\|_1 + \|u^-\|_1 + \|u^0\|_1)$$
$$\leq \int_\Omega \left(\ell(x) + \frac{\delta\lambda_{k+1}}{2}\right)(u^+)^2 + C_3(\|u^+\|_1 + \|u^-\|_1 + \|u^0\|_1) +$$
$$C_4(\|u^+\|_1^{1+\alpha} + \|u^-\|_1^{1+\alpha} + \|u^0\|_1^{1+\alpha})$$

which yields

$$(2.9) \quad 2q(u^+) - \int_\Omega \ell(x)(u^+)^2 dx - 2q(u^-) \leq \frac{\delta}{2}\lambda_{k+1}\|u^+\|_0^2 + C_3(\|u^+\|_1 + \|u^-\|_1 + \|u^0\|_1)$$
$$+ C_4(\|u^+\|_1^{1+\alpha} + \|u^-\|_1^{1+\alpha} + \|u^0\|_1^{1+\alpha}).$$

From lemma 2.1, (2.1) and (2.2), inequality (2.9) becomes

$$(2.10) \quad 2\delta\|u^+\|_1^2 + \left(\frac{\lambda_k}{\lambda_{k-1}} - 1\right)\|u^-\|_1^2 \leq \frac{\delta}{2}\|u^+\|^2 + C_3(\|u^+\|_1 + \|u^-\|_1 + \|u^0\|_1) +$$
$$C_4(\|u^+\|_1^{1+\alpha} + \|u^-\|_1^{1+\alpha} + \|u^0\|_1^{1+\alpha}).$$

Thus, using Young inequality in the right side of (2.10), we obtain

$$2\delta\|u^+\|_1^2 + \left(\frac{\lambda_k}{\lambda_{k-1}} - 1\right)\|u^-\|_1^2 \leq \delta\|u^+\|_1^2 + \frac{1}{2}\left(\frac{\lambda_k}{\lambda_{k-1}} - 1\right)\|u^-\|_1^2 + C_5(\|u^0\|_1^{1+\alpha} + 1)$$

hence

$$(2.11) \qquad \|u^+\|^2 + \|u^-\|^2 \leq C_6(\|u^0\|^{1+\alpha} + 1)$$

where C_i are positive constants non depending on u. It first follows from (2.11) that

$$|\|u_n\|_1 - \|u_n^0\|_1| \leq \|u_n - u_n^0\|_1 = \|u_n^+ + u_n^-\|_1 \leq [C_6(\|u_n^0\|_1^{1+\alpha} + 1)]^{1/2}$$

and so $\|u_n^0\|/\|u_n\| \to 1$, which implies

$$(2.12) \qquad \|u_n^0\| \to +\infty$$

Using now (2.11) in (2.8) yields

$$\Phi(u_n) \leq \|u_n^0\|_1^{1+\alpha} \left[C_7\left(1 + \frac{1}{\|u^0\|_1^{1+\alpha}}\right) - \frac{1}{\|u_n^0\|_1^{1+\alpha}}\int_\Omega F\left(x, \frac{u_n^0}{2}\right) dx\right]$$

where, by (2.12) and (f$_3$), the expression in square brackets tends to $-\infty$. Thus, we have reached a contradiction with $|\phi(u_n)| \leq C$. The proof when $k = 1$ is similar (and simpler). Q.E.D.

Proof of Theorem 1.1: Since the functional Φ is weakly lower semicontinuous and $\Phi|_{E^+}$ is coercive by lemma 2.3(ii), the infimum $b = Inf_{E^+}\Phi > -\infty$ is attained. For any given $a < b$, lemma 2.3(i) yields the existence of some $R > 0$ such that $\Phi(v) \leq a$ for all $v \in E^- \oplus E^0$ with $\|v\|_1 \geq R$. Thus, since Φ satisfies Palais-Smale condition, the saddle point theorem of Rabinowitz [?] insures the existence of a critical point $u_0 \in H_0^1(\Omega)$ of Φ with $\Phi(u_0) \geq b$. Q.E.D.

Corollaries and variants

We start by considering some more explicit conditions on F which guarantee the validity of assumption (f$_3$) in Theorem 1.1.

Corollary 3.1. *Assume* (f$_0$), (f$_1$), (f$_2$) *and*
(f$_4$) *for some β with $1 + \alpha < \beta \leq 2$*

$$\int_{v>0} F_+(x)|v|^\beta dx + \int_{v<0} F_-(x)|v|^\beta dx > 0 \quad \forall v \in E^0 = E(\lambda_k), v \neq 0,$$

where $F_\pm(x)$ denotes the nonnegative functions $F_\pm(x) = \liminf_{s \to \pm\infty} \frac{F(x,s)}{|s|^\beta}$ (these inferior limits being uniform with respect to x). Then, for any $h \in L^2(\Omega)$, problem (1.1) admits at least one solution.

Proof: It suffices to show that (f$_3$) holds. We will consider the functional $\varphi : E^0 \to \mathbb{R}^+$ defined by

$$\varphi(v) = \int_{v>0} F_+(x)|v|^\beta dx + \int_{v<0} F_-(x)|v|^\beta dx$$

and first prove that φ is lower semicontinuous. For that purpose, let $(v_n) \subset E$ be a sequence such that $v_n \to v$ in E^0. Passing to a subsequence if necessary, we can assume that $v_n \to v$ a.e in Ω. If $v \equiv 0$, then the inequality $\varphi(v) \leq \liminf \varphi(v_n)$ trivially holds since $\varphi(v) = 0$ and $\varphi(v_n) \geq 0$. Let us thus assume $v \not\equiv 0$, which implies, by the unique continuation property, that meas $\{x/v(x) = 0\} = 0$. Denoting by χ_n and χ the characteristic functions of the sets $\{x/v_n \geq 0\}$ and $\{x/v(x) \geq 0\}$ respectively, we then have $\chi_n \to \chi$ a.e in Ω. Writing $\varphi(v_n)$ as

$$\varphi(v_n) = \int_\Omega \chi_n F_+ |v_n|^\beta dx + \int_\Omega (1 - \chi_n) F_- |v_n|^\beta dx,$$

one deduces the desired lower-semicontinuity from Fatou's lemma.

Since, by (3.1), $\varphi > 0$ on the compact subset $S = \{v \in E^0/\|v\|_1 = 1\}$, the above lower semicontinuity implies the existence of $\varepsilon > 0$ such that

$$\varphi\left(\frac{v}{\|v\|_1}\right) = \int_{v>0} F_+ \frac{|v|^\beta}{\|v\|_1^\beta} dx + \int_{v<0} F_- \frac{|v|^\beta}{\|v\|_1^\beta} dx \geq \varepsilon$$

for all nonzero v in E^0. Let $\delta = \frac{\varepsilon}{2} \frac{\sqrt{\lambda_1}}{|\Omega|^{1-\frac{\beta}{2}}}$, where $|\Omega|$ denotes the measure of Ω. By the uniformity of the limits which define $F_\pm(x)$ and by (f_0)-(f_2), there exists $b_\delta(x) \in L^1(\Omega)$ such that

$$F(x,s) \geq (F_+(x) - \delta)|s|^\beta - b_\delta(x) \quad s \geq 0,$$
$$F(x,s) \geq (F_-(x) - \delta)|s|^\beta - b_\delta(x) \quad s \leq 0.$$

Consequently, for $v \in E^0$,

$$\int_\Omega F(x,v)dx \geq \int_{v>0} F_+(x)|v|^\beta dx + \int_{v<0} F_-(x)|v|^\beta dx - \delta \int_\Omega |v|^\beta dx - \|b_\delta\|_{L^1}$$
$$\geq \int_{v>0} F_+(x)|v|^\beta dx + \int_{v<0} F_-(x)|v|^\beta dx - \frac{\varepsilon}{2}\|v\|_1^\beta - C,$$

and so

$$\frac{1}{\|v\|_1^{1+\alpha}} \int_\Omega F(x,v)dx \geq \|v\|_1^{\beta-1-\alpha} \varphi\left(\frac{v}{\|v\|_1}\right) - \frac{\varepsilon}{2}\|v\|_1^{\beta-1-\alpha} - \frac{C}{\|v\|_1^{1+\alpha}}$$
$$\geq \frac{\varepsilon}{2}\|v\|_1^{\beta-1-\alpha} - \frac{C}{\|v\|_1^{1+\alpha}}.$$

Since $1 + \alpha < \beta$, the above estimate shows that $\frac{1}{\|v\|_1^{1+\alpha}} \int_\Omega F(x,v)dx \to +\infty$ as $\|v\|_1 \to +\infty$, $v \in E^0$, i.e (f_3). Q.E.D.

Corollary 3.2. Assume (f_0), (f_1), (f_2) and

(f_5) $$\lim_{|s| \to +\infty} \frac{F(x,s)}{|s|^{1+\alpha}} = +\infty, \text{ uniformly for a.e } x \in \Omega.$$

Then, for any $h \in L^2(\Omega)$, problem (1.1) admits at least one solution.

Proof: As in lemma C.15 of [13], one shows that (f_5) yields (f_3). Q.E.D.

We now turn to a "dual" version of Theorem 1.1.

Theorem 3.3. Assume (f_0),

(f_1') $$\lambda_{k-1} - \lambda_k \leq k(x) \leq \liminf_{s \to \pm\infty} \frac{f(x,s)}{s}, k \geq 2$$

uniformly in x, with meas $\{x \in \Omega/\lambda_{k-1} - \lambda_k < k(x)\} > 0$,

(f_2') $$f(x,s) \leq as^\alpha + b(x) \quad x \in \Omega, s \geq 0,$$
$$f(x,s) \geq -as^\alpha - b(x) \quad x \in \Omega, s \leq 0,$$

where $b(x) \in L^2(\Omega)$, and $0 \leq \alpha < 1$,

(f_3') $\qquad \dfrac{1}{\|u^0\|_1^{1+\alpha}} \displaystyle\int_\Omega F(x, u^0)dx \to -\infty$ as $\|u^0\|_1 \to +\infty, u^0 \in E(\lambda_k)$.

Then, for any $h \in L^2(\Omega)$, problem (1.1) admits at least one solution.

The proof of theorem 3.3 is similar to that of Theorem 1.1. Corollaries analogous to corollaries 3.1 and 3.2 above can of course also be derived from Theorem 3.3.

Examples

Example 4.1. Let $\gamma \in]0, \lambda_{k+1} - \lambda_k[$, $a_0' = 0$, $a_n = 2^{2n} - \dfrac{1}{2^{3n+1}}$, $a_n' = 2^{2n} + \dfrac{1}{2^{3n+1}}, n \geq 1$. Let $f : \mathbf{R} \to \mathbf{R}$ be the odd continuous function defined in the following way: put

$$f(s) = \begin{cases} \gamma s & \text{on } [a_n', a_n], n \geq 1, \\ -2^n \gamma & \text{at each point } 2^{2n}, \end{cases}$$

and connect the values at the three successive points a_n, 2^{2n}, a_n' in a piecewise linear manner. Remark that

$$0 = \liminf_{s \to \pm\infty} \dfrac{f(s)}{s} < \limsup_{s \to \pm\infty} \dfrac{f(s)}{s} = \gamma < \lambda_{k+1} - \lambda_k.$$

A simple estimate of the primitive $F(s) = \int_0^s f(t)dt$ gives

$$\dfrac{\gamma s^2}{2} - 1 \leq F(s) \leq \gamma \dfrac{s^2}{2}$$

and so $\lim_{|s| \to +\infty} \dfrac{F(s)}{|s|\sqrt{|s|}} = +\infty$. Corollary 3.2 thus implies that problem (1.1) is solvable for any $h \in L^2(\Omega)$. Note that the result of [12] does not apply here.

Example 4.2. Let $\tilde{f} : \mathbf{R} \to \mathbf{R}$ be the odd function defined by

$$\tilde{f} = f^+ \text{ on } \mathbf{R}^+,$$

where f is as in example 4.1. Remark that by corollary 3.2, problem (1.1) with f substituted by $\tilde{f}$ admits a solution for any $h \in L^2(\Omega)$. We wish to make a comparison with a recent work of Fabry and Fonda [5] in which the solvability of problem (1.1) is derived under the assumption

(4.1) $\qquad \begin{cases} sign(s)f(x, s) \geq -b(x), \\ sign(s)[(\lambda_{k+1} - \lambda_k)s - f(x, s)] \geq -b(x), \end{cases}$

where $b(x) \in L^2(\Omega)$, and the following Landesman-Lazer conditions:

(4.2) $\displaystyle\int_{v > 0} \left(\liminf_{s \to +\infty} f(x, s) + h(x)\right) v(x)dx + \displaystyle\int_{v < 0} \left(\limsup_{s \to -\infty} f(x, s) + h(x)\right) v(x)dx > 0$

for all $v \in E^0 \setminus \{0\}$,

(4.3)
$$\int_{w>0} \liminf_{s\to+\infty} \left[(\lambda_{k+1} - \lambda_k)s - f(x,s) - h(x)\right] w\, dx +$$
$$\int_{w<0} \limsup_{s\to-\infty} \left[(\lambda_{k+1} - \lambda_k)s - f(x,s) - h(x)\right] w\, dx > 0$$

for all $w \in E(\lambda_{k+1}) \setminus \{0\}$.

It is clear that $\tilde{f}$ satisfies (4.1) and (4.3). However, for any h, (4.2) does not hold since $\liminf_{s\to+\infty} \tilde{f}(s) = \limsup_{s\to-\infty} \tilde{f}(s) = 0$.

Example 4.3. Let us define $a'_0 = 0$, $a_n = 2^{4n} - \frac{1}{2^{4n}}$, $a'_n = 2^{4n} + \frac{1}{2^{4n}}$, $b_n = \frac{a'_{n-1} + a_n}{2}$, $b_n^- = b_n - \frac{1}{2^{2n}}$ and $b_n^+ = b_n + \frac{1}{2^{2n}}$, $n \geq 1$. Let $f : \mathbf{R} \to \mathbf{R}$ be an odd function defined in the following way: put

$$f(s) = \begin{cases} s^{\frac{1}{4}} & \text{on } [a'_{n-1}, b_n^-] \text{ and } [b_n^+, a_n], n \geq 1 \\ 2^{3n} & \text{at each point } 2^{4n}, n \geq 1 \\ -(b_n)^{\frac{1}{8}} & \text{at each point } b_n, n \geq 1 \end{cases}$$

and connect the values at the three successive points b_n^-, b_n, b_n^+ in a piecewise linear manner and similarly at the three successive points a_n, 2^{4n}, a'_n. It is clear that f satisfies (f_0), (f_1) and (f_2). A simple estimate of $F(s) = \int_0^s f(t)\,dt$ gives

$$\frac{4}{5}|s|^{\frac{5}{4}} - 2 \leq F(s) \leq \frac{4}{5}|s|^{\frac{5}{4}} + 1.$$

Consequently $\lim_{|s|\to+\infty} \frac{F(s)}{|s|^{9/8}}$ and so, by corollary 3.2 (with $\alpha = \frac{1}{8}$), problem (1.1) is solvable for any $h \in L^2(\Omega)$. This kind of nonlinearity is not covered by Silva's result [13] mentionned in the introduction. Indeed f satisfies

$$|f(s)| \leq |s|^{\frac{3}{4}} + 1$$

but $\frac{1}{\|u^0\|_1^{6/4}} \int_\Omega F(u^0)\,dx \to 0$ as $\|u^0\|_1 \to +\infty$, $u^0 \in E^0$.

References

[1] D.G.Costa, *A note on unbounded perturbations of linear resonant problems*, Preprint Univ. Brasilia, Trabalho de Matematica n°245 (1989), 1-8.

[2] D.G.Costa, A.Oliveira, *Existence of solution for a class of semilinear elliptic problems at double resonance*, Boll. Soc. Bras. Mat., (19)(1) 1988, 21-37.

[3] D.De Figueiredo, J.P.Gossez, *Conditions de non résonance pour certains problèmes elliptiques semi-linéaires*, C.R. Acad. Sc. Paris, 302 (1986), 543-545.

[4] C.L.Dolph, *Nonlinear integral equations of the Hammerstein type*, Trans. Amer. Math. Soc., 66 (1949), 289-307.

[5] C.Fabry, A.Fonda, *Nonlinear equations at resonance and generalised eigenvalue problems*, Sém. Math. 173, Univ. Catholique de Louvain (1990).

[6] A.Fonda, D.Lupo, *Periodic solutions of second order ordinary differential equations*, Boll. U.M.I. (7) 3.A (1989), 291-299.

[7] J.P.Gossez, *Nonresonance near the first eigenvalue of a second order elliptic problem*, Proc. E.L.A.M., Springer Lectures Notes ,1324 (1988),97-104.

[8] J.P.Gossez, P.Omari, *Periodic solutions of a second order ordinary differential equation: a necessary and sufficient condition for nonresonance*, J. Diff. Equat., 94(1994),67-82.

[9] J.Mawhin, J.Ward, M.Willem, *Variational methods and semilinear elliptic equations*, Arch. Rat. Mech. Anal., 95 (1986), 269-277

[10] M.Moussaoui, *Questions d'existence dans les problèmes semi-linéaires elliptiques*, Thèse doct., Université Libre de Bruxelles (1991).

[11] P.H.Rabinowitz, *Minimax methods in critical point theory with applications to differential equations*, C.B.M.S. Reg. Conf. 65, Amer. Math. Soc. (1986).

[12] M.Ramos, *Remarks on resonance problems with unbounded perturbations*, Séminaire de Mathématique 172, Univ. Catholique de Louvain (1990).

[13] E.A. Silva, *Critical point theorems and applications to differential equations*, Ph. D. thesis, Univ. Wisconsin, Madison, 1988.

ADDRESS

Jean-Pierre GOSSEZ, Département de Mathématique, Université Libre de Bruxelles, 1050 Bruxelles, Belgique.

Mimoun MOUSSAOUI, Département de Mathématique, Université Mohammed I, Oujda, Maroc.

C A STUART
Cylindrical TE and TM modes in a self-focusing dielectric

INTRODUCTION

This lecture reviews some recent work concerning the existence and behaviour of self-guided beams of light in a nonlinear medium. In order to model propagation in a optical fibre we seek beams having a cylindrical symmetry. The propagation of light is governed by Maxwell's equations together with a nonlinear relationship between the electric field and the displacement field. For a guided beam the light must remain concentrated near the axis of propagation and the amplitudes of the electro-magnetic fields must decay to zero far from this axis.

Mathematically, the problem can be formulated by seeking solutions of Maxwell's equations which have an appropriate cylindrical symmetry and which satisfy certain boundary conditions. There are two types of solution for which Maxwell's equations can be reduced, without approximation, to a much simpler form which is amenable to a rigorous analysis. These are the TE (transverse electric) and the TM (transverse magnetic) field modes described below. In this lecture some of the results concerning these special modes are summarized. They involve the analysis of a boundary-value problem for a second order differential equation on the interval $(0, \infty)$. The equation is of semilinear type in the case of TE-modes, whereas for TM-modes it has a more complicated quasilinear form.

The presentation is organized as follows.

§1. Mathematical formulation of the guidance problem.
§2. Analysis of TE-modes.
§3. Analysis of TM-modes.
§4. Related problems.

1. MATHEMATICAL FORMULATION OF THE GUIDANCE PROBLEM

In a dielectric medium Maxwell's equations can be written as [1],

$$\nabla \wedge E = -\frac{1}{c} \partial_t B \, , \, \nabla \wedge H = \frac{1}{c} \partial_t D,$$

$$\nabla \cdot D = 0, \nabla \cdot B = 0 \tag{ME}$$

where $c > 0$ is the speed of light in a vacuum and the electro-magnetic fields are functions of space, (r, θ, z), and time, t.

Here (r, θ, z) denote cylindrical polar co-ordinates and the usual orthonormal basis associated with this system is denoted by

$$i_r = \begin{pmatrix} \cos\theta \\ \sin\theta \\ 0 \end{pmatrix}, \quad i_\theta = \begin{pmatrix} -\sin\theta \\ \cos\theta \\ 0 \end{pmatrix}, \quad i_z = \begin{pmatrix} 0 \\ 0 \\ 1 \end{pmatrix}$$

A field $F: \Re^4 \to \Re^3$ can be resolved into orthogonal components (F_r, F_θ, F_z) by setting

$$F_\alpha(r, \theta, z, t) = F(r, \theta, z, t) \cdot i_\alpha \quad \text{for } \alpha = r, \theta, z. \tag{1.1}$$

We seek solutions of Maxwell's equations which have the form of monochromatic cylindrical modes propagating in the direction of the z-axis. More precisely, each field F should have the form

$$F_\alpha(r, \theta, z, t) = \begin{cases} f_\alpha(r) \cos(kz - \omega t) & \text{for } \alpha = r, \theta \\ f_\alpha(r) \sin(kz - \omega t) & \text{for } \alpha = z \end{cases} \tag{1.2}$$

where $f_\alpha : [0, \infty) \to \Re$ is a scalar function and the positive constants k and ω give the wavelength, $\frac{2\pi}{k}$, and the frequency ω of the associated beam.

For such modes the behaviour of the medium is characterized by the following constitutive relations[2,3,4],

$$\begin{aligned} B(r, \theta, z, t) &= H(r, \theta, z, t) \\ D(r, \theta, z, t) &= \varepsilon\left(\frac{1}{2}\left[e_r^2 + e_\theta^2 + e_z^2\right](r)\right) E(r, \theta, z, t) \end{aligned} \tag{CR}$$

For an isotropic medium, $\varepsilon : [0, \infty) \to (0, \infty)$ is a scalar function called the dielectric response of the medium. The quantity $\frac{1}{2}\left[e_r^2 + e_\theta^2 + e_z^2\right](r)$ is the time-average of $|E(r, \theta, z, t)|^2$ since E has the form (1.2).

It is physically reasonable to assume that ε has the following properties.

(A) $\varepsilon \in C([0, \infty)) \cap C^1((0, \infty))$ with $\varepsilon'(s) \geq 0 \quad \forall s > 0$, $\varepsilon(0) > 0$, $\varepsilon(\infty) = \lim_{s \to \infty} \varepsilon(s) < \infty$ and $\lim_{s \to 0} s\varepsilon'(s) = 0$. Furthermore, there exist positive constants L and σ such that $\lim_{s \to 0} \frac{\varepsilon(s) - \varepsilon(0)}{s^\sigma} = L$.

Since ε is increasing the dielectric response is of the type called self-focusing. The fact that $\varepsilon(\infty) < \infty$ means that this response saturates as the field strength becomes infinite. For a large class of materials $\sigma = 1$, but other values do occur for some materials, [5,6,7].

Finally we introduce the conditions on the $e - m$ fields corresponding to the requirement that they represent a confined beam of light.

Guidance Conditions

(i) $\int_0^{2\pi} \int_0^\infty [E \cdot D + H \cdot B]\, r\, dr\, d\theta < \infty$

(ii) $\lim_{r \to \infty} F(r, \theta, z, t) = 0$ where $F = E, D, B$ and H.

The first condition asserts that the total $e - m$ energy in planes transverse to the direction of propagation is finite and the second one means that the fields decay to zero far from the axis of propagation.

In what follows we seek solutions of (ME) and (CR) in the form (1.2) that satisfy these guidance conditions. The strength of these guided beams is measured by the time-average of the energy flux

$$\int_0^{2\pi} \int_0^\infty c(E \wedge H) \cdot i_z \, r\, dr\, d\theta$$

across planes transverse to the direction of propagation. Using (1.2) this quantity becomes

$$P = \pi c \int_0^\infty (e_r h_\theta - e_\theta h_r)\, r\, dr \qquad (1.3)$$

and it is referred to as the power of the beam.

In the following discussion, the frequency ω is fixed and we consider the relation between the power, P, and the wavelength, $\frac{2\pi}{k}$, for guided modes. The results are expressed most conveniently using the variable $\mu = \left(\frac{\omega}{ck}\right)^2$ instead of $\frac{2\pi}{k}$.

Both TE and TM modes have the following common features provided that the dielectric response satisfies (A).

(i) For all guided beams $\frac{1}{\varepsilon(\infty)} < \mu < \frac{1}{\varepsilon(0)}$.

(ii) For all guided beams, the amplitudes of the $e - m$ fields decay exponentially to zero as $r \to \infty$.

(iii) Guided beams exist at sufficiently high powers and, as $P \to \infty$, we have $\mu \to \frac{1}{\varepsilon(\infty)}$.

(iv) If $\sigma \geq 1$ in (A), there are no guided beams with power below a certain threshold, $P_o > 0$.

(v) If $0 < \sigma < 1$ in (A), guided beams exist at arbitrarily small powers and, as $P \to 0$ we have $\mu \to \frac{1}{\varepsilon(0)}$.

Our formulation based on the unknown e_θ for TE-modes is standard. On the other hand, due to the form of the constitutive relation (CR), the basic unknown in the study of cylindrical TM-modes is usually taken to be the couple (e_r, e_z). (See [15], for example.) Setting $e_r = \varphi$ and $e_z = k\psi$, (ME) with (CR) reduce to the following system of equations,

$$\{\varphi(r) + \psi'(r)\} = \mu\varepsilon \left(\frac{1}{2}\left[\varphi(r)^2 + k^2\psi(r)^2\right]\right)\varphi(r)$$

$$-\frac{1}{r}\left[r\{\varphi(r) + \psi'(r)\}\right]' = \mu\varepsilon \left(\frac{1}{2}\left[\varphi(r)^2 + k^2\psi(r)^2\right]\right)\psi(r)$$

where $\mu = \left(\frac{\omega}{ck}\right)^2$. However in [9] we have shown that by inverting (CR) (see (CR)* in Section 3) we are able to use the scalar function h_θ as the basic unknown for the TM-modes. In this way we obtain a formulation in the TM case which closely resembles that for TE-modes. In particular the equations (2.2) and (3.2) have exactly the same linearizations at $v \equiv 0$, namely the Bessel equation

$$\left[v'(r) + \frac{v(r)}{r}\right]' + \left\{\frac{\omega^2}{c^2}\varepsilon(0) - k^2\right\}v(r) = 0$$

since $\gamma(0) = \frac{1}{\varepsilon(0)}$. The low-power modes mentioned in (v) bifurcate from the infimum of the essential spectrum of this linear eigenvalue problem.

2. CYLINDRICAL TE-MODES

We seek solutions for which the electric field E has the additional properties that $e_r \equiv e_z \equiv 0$.
Setting $e_\theta = v$, we have

$$E(r, \theta, z, t) = v(r)\cos(kz - \omega t) i_\theta \tag{2.1}$$

In [8] we have shown that (ME) together with (CR) are satisfied in the region $r > 0$ provided that $v \in C^2((0, \infty))$ and satisfies the equation

$$\left[v'(r) + \frac{v(r)}{r}\right]' + \frac{\omega^2}{c^2}\varepsilon\left(\frac{1}{2}v(r)^2\right)v(r) - k^2 v(r) = 0. \tag{2.2}$$

From a solution v of this equation the displacement and magnetic fields are defined by

$$D(r,\theta,z,t) = \varepsilon\left(\frac{1}{2}v(r)^2\right) E(r,\theta,z,t),$$

$$h_r(r) = -\frac{ck}{\omega}v(r), \quad h_\theta(r) = 0, \quad h_z(r) = \frac{c}{\omega}\left[v(r) + \frac{v(r)}{r}\right] \text{ and}$$

$$B(r,\theta,z,t) = H(r,\theta,z,t).$$

To extend these fields smoothly onto the axis $r = 0$, we require that

$$\lim_{r \to 0} v(r) = \lim_{r \to 0}\left[v'(r) + \frac{v(r)}{r}\right]' = 0 \text{ and that } \lim_{r \to 0} v'(r) \text{ exists.}$$

Furthermore the guidance conditions reduce to

(i) $\int_0^\infty \left\{\left[\frac{c^2 k^2}{\omega^2} + \varepsilon\left(\frac{1}{2}v(r)^2\right)\right] v(r)^2 + \frac{c^2}{\omega^2}\left[v(r) + \frac{v(r)}{r}\right]^2\right\} r\, dr < \infty$

and

(ii) $\lim_{r \to \infty} v(r) = \lim_{r \to \infty} v'(r) = 0$.

The power (1.3) of such a guided beam is given by

$$P = \frac{\pi c^2 k}{\omega} \int_0^\infty v(r)^2 \, r\, dr.$$

For the analysis of this problem it is convenient to introduce the new variables defined by

$$u(r) = r^{1/2} v(r) \quad \text{and} \quad \eta = -k^2$$

Then we have shown [8] that the guidance problem is equivalent to finding $(\eta, u) \in (-\infty, 0) \times H_0^1(0, \infty)$ such that

$$u \not\equiv 0 \text{ and } J'(u) = \eta u \tag{GE}$$

where $J : H_0^1(0, \infty) \to \Re$ is defined by

$$J(u) = \int_0^\infty \frac{1}{2}\left[u'(r) + \frac{u(r)}{2r}\right]^2 - r\, E\left(\frac{1}{2r}u(r)^2\right) dr$$

with $E(s) = \left(\frac{\omega}{c}\right)^2 \int_0^s \varepsilon(t) dt$.

Using Hardy's inequality it is easy to check that J is C^1 and, for any solution (η, u) of (GE), we see that $u \in C^2((0, \infty))$ and $-\frac{\omega^2}{c^2} \varepsilon(\infty) < \eta < 2J(u)$ since $E(s) < \left(\frac{\omega}{c}\right)^2 \varepsilon(s) s \ \forall s > 0$.

One way of obtaining solutions of (GE) is to solve the following minimization problem.

Given $d > 0$, set

$$S(d) = \left\{ u \in H_0^1(0, \infty) : \int_0^\infty u(r)^2 dr = d^2 \right\} \quad \text{and} \quad \text{(ME)}$$

$$m(d) = \inf \{ J(u) : u \in S(d) \}.$$

Then find $u \in S(d)$ such that $J(u) = m(d)$.

Clearly if u is a solution of (ME) then there is a Lagrange multiplier η such that (η, u) satisfies (GE) provided that $\eta < 0$. However (ME) does not always have a solution. Indeed we have shown in Theorem 4.6(ii) of [8] that, if $\sigma \geq 1$ in (A), then there exists $d_1 > 0$ such that $\int_0^\infty u(r)^2 dr \geq d_1^2$ for all solutions (η, u) of (GE).

Consequently (ME) has no solution for $d \in (0, d_1)$. This happens because in such cases minimizing sequences on $S(d)$ converge weakly to zero, and this in turn implies that $m(d) \geq -\frac{1}{2} \frac{\omega^2}{c^2} \varepsilon(0) d^2$. But, for any $\sigma > 0$ in (A),

$$\lim_{d \to \infty} \frac{m(d)}{d^2} \leq -\frac{1}{2} \frac{\omega^2}{c^2} \varepsilon(\infty) < -\frac{1}{2} \frac{\omega^2}{c^2} \varepsilon(0)$$

and so $\exists d_o \geq 0$ such that

$$m(d) < -\frac{1}{2} \frac{\omega^2}{c^2} \varepsilon(0) d^2 \text{ for all } d > d_o.$$

Furthermore, if $0 < \sigma < 1$ in (A), we can set $d_o = 0$. Hence for $d > d_o$, we are able to show that (ME) has a solution, u_d, with $u_d > 0$ on $(0, \infty)$. The corresponding Lagrange multiplier satisfies the following inequalities,

$$-\frac{\omega^2}{c^2} \varepsilon(\infty) < \eta_d < 2m(d)/d^2 < -\frac{\omega^2}{c^2} \varepsilon(0)$$

where $\lim_{d \to \infty} \eta_d = -\frac{\omega^2}{c^2} \varepsilon(\infty)$ and, if $0 < \sigma < 1$ in (A), $\lim_{d \to 0} \eta_d = -\frac{\omega^2}{c^2} \varepsilon(0)$.

In proving these results in [8] we found it convenient to replace J by

$$\hat{J}(u) = J(u) + \frac{1}{2} \left(\frac{\omega}{c}\right)^2 \varepsilon(0) \int_0^\infty u(r)^2 dr.$$

The main steps in proving that (ME) has a solution are
(a) to show that $\hat{J} : H_o^1(0, \infty) \to \Re$ is weakly sequentially lower semicontinuous, and
(b) to show that $\hat{m}(d) < 0$ where

$$\hat{m}(d) = \inf \left\{ \hat{J}(u) : u \in S(d) \right\} = m(d) + \frac{1}{2}\left(\frac{\omega}{c}\right)^2 \varepsilon(0) d^2.$$

It is in the analysis of part (b) that the difference between the cases $0 < \sigma < 1$ and $\sigma \geq 1$ occurs.

3. CYLINDRICAL TM-MODES

We seek solutions for which the magnetic fields have the additional properties that $h_r \equiv b_r \equiv h_z \equiv b_z \equiv 0$.
Setting $h_\theta(r) = b_\theta(r) = \frac{\omega}{c} v(r)$, this means that

$$H(r, \theta, z, t) = B(r, \theta, z, t) = \frac{\omega}{c} v(r) \cos(kz - \omega t) \, i_\theta. \qquad (3.1)$$

Our initial aim is to reduce Maxwell's equations (ME) with the constitutive relations (CR) to an equation for v. The first step involves an inversion the constitutive relation. This is done as follows.
Supposing that ε satisfies (A), we set

$$f(s) = \varepsilon(s)^2 s \text{ for } s \geq 0$$

and then define a new function γ by

$$\gamma(\tau) = \begin{cases} 1/\varepsilon(0) \text{ for } \tau = 0 \\ \sqrt{f^{-1}(\tau)/\tau} \text{ for } \tau > 0 \end{cases}.$$

It is shown in [9] that γ is well-defined and has the following properties.

(H) $\gamma \in C\left([0, \infty)\right) \cap C^1\left((0, \infty)\right)$ with $\gamma'(\tau) \leq 0 < \gamma(\tau) + 2\tau\gamma'(\tau)$ for all $\tau > 0$, $\gamma(\infty) = \lim_{\tau \to \infty} \gamma(\tau) = 1/\varepsilon(\infty) > 0$ and $\lim_{\tau \to 0} \tau\gamma'(\tau) = 0$. Furthermore, $\lim_{\tau \to 0} \frac{\gamma(\tau) - \gamma(0)}{\tau^\sigma} = -K$ where $K = L/\varepsilon(0)^{2(1+\sigma)}$.

According to Proposition 1.1 of [9] the constitutive relationship (CR) is equivalent to

$$H(r, \theta, z, t) = B(r, \theta, z, t) \qquad (CR^*)$$
$$E(r, \theta, z, t) = \gamma\left(\frac{1}{2}\left[d_r^2 + d_\theta^2 + d_z^2\right](r)\right) D(r, \theta, z, t).$$

173

Using this it can be shown [9] that Maxwell's equations with the constitutive relation (CR) are satisfied in the region $r > 0$ provided that $v \in C^2((0,\infty))$ and v satisfies the differential equation

$$\left\{\gamma\left(\frac{1}{2}\left[k^2v(r)^2 + w(r)^2\right]\right)w(r)\right\}' - k^2\gamma\left(\frac{1}{2}\left[k^2v(r)^2 + w(r)^2\right]\right)v(r) + \frac{\omega^2}{c^2}v(r) = 0 \quad (3.2)$$

where $w(r) = v'(r) + \frac{v(r)}{r}$.

Eliminating w, we see that this is a single second order quasilinear differential equation for v.

From a solution v of this equation the electric fields are defined by

$$d_r(r) = kv(r), \quad d_\theta(r) \equiv 0, \quad d_z(r) = -w(r) \quad \text{and}$$

$$E(r,\theta,z,t) = \gamma\left(\frac{1}{2}\left[k^2v(r)^2 + w(r)^2\right]\right) D(r,\theta,z,t).$$

In order to extend these fields onto the axis $r = 0$ we require that

$$\lim_{r\to 0} v(r) = 0 \text{ and } \lim_{r\to 0} v'(r) = 0.$$

(See Proposition 5.1 of [9].)
The guidance conditions reduce to

(i) $\int_0^\infty \left\{\frac{\omega^2}{c^2}v(r) + \gamma\left(\frac{1}{2}[k^2v(r)^2 + w(r)^2]\right)[k^2v(r)^2 + w(r)^2]\right\} r\, dr < \infty$

and

(ii) $\lim_{r\to\infty} v(r) = 0$ and $\lim_{r\to\infty} v'(r) = 0$.

The power (1.3) of such guided beams is given by

$$P = \pi\omega k^2 \int_0^\infty \gamma\left(\frac{1}{2}\left[k^2v(r)^2 + w(r)^2\right]\right) v(r)^2 r\, dr.$$

For the analysis of this problem it is convenient to introduce new variables as follows,

$$u(r) = k\sqrt{r}\, v(r/k) \text{ and } \mu = \left(\frac{\omega}{ck}\right)^2.$$

Then as is shown in [10] the guidance problem is equivalent to finding $(\mu, u) \in (0,\infty) \times H_0^1((0,\infty))$ such that

$$u \not\equiv 0 \text{ and } j(u) = \mu u \quad \text{(GM)}$$

where $j : H_0^1(0,\infty) \to \Re$ is defined by

$$j(u) = \int_0^\infty \Gamma\left(\frac{1}{2r}\left[u(r)^2 + (Tu(r))^2\right]\right) r \, dr$$

with $Tu(r) = u'(r) + \frac{u(r)}{2r}$ and $\Gamma(\tau) = \int_0^\tau \gamma(t)dt$
By (H), $\Gamma \in C^1([0,\infty))$ and $0 < \gamma(\infty)\tau \leq \Gamma(\tau) \leq \gamma(0)\tau \quad \forall \tau > 0$.

Hardy's inequality implies that $T : H_0^1(0,\infty) \to L^2(0,\infty)$ is a bounded linear operator. Hence j is C^1 and for any solution of (GM) we find that $\mu > \gamma(\infty)$. By applying regularity theorems due to Tonelli we have shown in Theorem 3.3 of [10] that $u \in C^2((0,\infty))$ whenever (μ, u) is a solution of (GM).

One way of obtaining solutions of (GM) is to solve the following minimization problem.

Given $d > 0$, set

$$S(d) = \left\{u \in H_0^1(0,\infty) : \int_0^\infty u(r)^2 dr = d^2\right\} \text{ and} \quad \text{(MM)}$$
$$M(d) = \inf\{j(u) : u \in S(d)\}.$$

Then find $u \in S(d)$ such that $j(u) = M(d)$.

Clearly if u is a solution of (MM) then there exists a Lagrange multiplier μ such that (μ, u) satisfies (GM). However (MM) does not always have a solution.

Indeed we have shown in Theorem 4.6 of [10] that if $\sigma \geq 1$ in (A) then there exists $d_1 > 0$ such that $\int_0^\infty u(r)^2 dr \geq d_1^2$ for all solutions (μ, u) of (GM). Consequently (MM) has no solution for $d \in (0, d_1)$. In such cases minimizing sequences on $S(d)$ converge weakly to zero. On the other hand, if a minimizing sequence on $S(d)$ converges weakly to zero then we find that $M(d) \geq \gamma(0)d^2/2$.

But for any $\sigma > 0$ in (A),

$$\lim_{d \to \infty} M(d)/d^2 \leq \gamma(\infty)/2 < \gamma(0)$$

and so there exists $d_0 \geq 0$ such that

$$M(d) < \gamma(0)d^2/2 \text{ for all } d > d_0.$$

Furthermore, if $0 < \sigma < 1$ in (A), we can set $d_0 = 0$.

In such cases, minimizing sequences on $S(d)$ cannot converge weakly to zero and we are able to show that (MM) has a solution, u_d, with $u_d > 0$ on $(0,\infty)$. The corresponding Lagrange multiplier satisfies the following inequalities

$$\gamma(\infty) < \mu_d < \gamma(0) \text{ for all } d > d_0$$

where
$$\lim_{d \to \infty} \mu_d = \gamma(\infty) \text{ and, if } 0 < \sigma < 1, \lim_{d \to 0} \mu_d = \gamma(0).$$

In proving these results in [10] we found it convenient to replace j by

$$\hat{j}(u) = j(u) - \frac{1}{2}\gamma(0)\int_0^\infty u(r)^2 dr.$$

The main steps in proving that (MM) has a solution are
(a) to show that $\hat{j}: H_0^1(0,\infty) \to \Re$ is weakly sequentially lower semicontinuous, and
(b) to show that $\hat{M}(d) < 0$ where

$$\hat{M}(d) = \inf\left\{\hat{j}(u) : u \in S(d)\right\} = M(d) - \frac{1}{2}\gamma(0)d^2.$$

It is in the analysis of part (b) that the difference between the cases $0 < \sigma < 1$ and $\sigma \geq 1$ occurs.

4. RELATED RESULTS

We end with some remarks about variants of the problems discussed in Sections 2 and 3.

Higher modes

In the variational principles (ME) and (MM) we have only considered the existence of fundamental TE and TM modes. However higher modes can also be found by studying all critical points of J and j on $S(d)$, rather than just looking for minima. This has been done by H.-J. Ruppen in a series of interesting papers [11,12].

Inhomogeneity of the medium

The assumption (A) requires the medium to be homogeneous. In the context of optical fibres it is natural to allow the material composition to vary with distance from the axis $r = 0$. This amounts to considering ε to be a function of two variables, $\varepsilon(r,s)$, where $s = \frac{1}{2}[e_r^2 + e_\theta^2 + e_z^2](r)$. The discussion of TE-modes in [8] covers this case. For TM-modes work in this direction in underway.

Defocusing media

The assumption (A) requires the medium to be self-focusing since $\varepsilon' \geq 0$. However some materials have a dielectric response for which $\varepsilon' \leq 0$. The existence of cylindrical guided modes has also been studied in such defocusing media, but it is essential to allow appropriate inhomogeneity of the fibre to counteract the dispersive effect of the nonlinearity. The case of cylindrical TE-modes is discussed in [13], and the study of TM-modes is in progress.

Planar waveguides

Analogous results are available for the case of planar rather than cylindrical symmetry. See [12] and [16] for the case of planar TE-modes. For planar TM-modes work is in progress.

REFERENCES

[1] Born, M. & Wolf, E.: Principles of Optics, fifth edition, Pergamon Press, Oxford, 1975.

[2] Akhmanov, R.-V., Khorklov, R.V. & Sukhorukov, A.P.: Self-focusing, self-defocusing and self-modulation of laser beams, in Laser Handbook, edited by F.T. Arecchi & E.O. Schulz Dubois, North-Holland, Amsterdam, 1972.

[3] Svelto, O.: Self-focusing, self-trapping and self-phase modulation of laser beams, in Progress in Optics, Vol. 12, editor E. Wolf, North-Holland, Amsterdam, 1974.

[4] Reintjes, J.F.: Nonlinear Optical Processes, in Encyclopedia of Physical Science and Technology, Vol. 9, Academic Press, New York.

[5] Stegeman, G.I., Ariyant, J., Seaton, C.I., Shen, T.-P. & Moloney, J.V.: Nonlinear thin-film guided waves in non-Kerr media, Appl. Phys. Lett. 47 (1985), 1254-1256.

[6] Mihalache, D., Bertolotti, M. & Sibilia, C.: Nonlinear wave propagation in planar structures, in Progress in Optics XXVII, edited by E. Wolf, Elsevier, 1989.

[7] Mathew, J.G.H., Lar, A.K., Heckenberg, N.R. & Galbraith, I.: Time resolved self-defocusing in InSb at room temperature, IEEE J. Quantum Elect. 21 (1985), 94-99.

[8] Stuart, C.A.: Self-trapping of an eletromagnetic field and bifurcation from the essential spectrum, Arch. Rational Mech. Anal. (1991), 65-96.

[9] Stuart, C.A.: Magnetic field wave equation for nonlinear optical waveguides, pre-print.

[10] Stuart, C.A.: Cylindrical TM-modes in a homogeneous self-focusing dielectric, pre-print.

[11] Ruppen, H.-J.: Multiple TE-modes for cylindrical, self-focusing waveguides, pre-print.

[12] Ruppen, H.-J.: Multiple TE-modes for planar, self-focusing waveguides, pre-print.

[13] John, O. and Stuart, C.A.: Guidance properties of a cylindrical defocusing wave-guide, to appear in Comm. Math. Univ. Carol.

[14] Chen, Y. and Snyder, A.W.: TM-type self-guided beams with circular cross-section, Electr. Lett., 27 (1991), 565-566.

[15] Chen, Y.: TE and TM families of self-trapped beams, IEEE, J. Quant. Electr., 27 (1991), 1236-1241.

[16] Stuart, C.A.: Guidance properties of nonlinear planar waveguides, Arch. Rational Mech. Anal., 125 (1993), 145-200.

Département de Mathématique - Ecole Polytechnique Fédérale - Lausanne - Suisse

J L VAZQUEZ
Entropy solutions and the uniqueness problem for nonlinear second-order elliptic equations

Introduction

The theory of *quasilinear elliptic equations* is concerned with equations of the form

(0.1) $$\operatorname{div} \mathcal{A}(x, u, \nabla u) = B(x, u, \nabla u),$$

where $\mathcal{A} = (A_i(x, u, \xi))$ is a vector-valued function defined in some open subset of $Q = \mathbf{R}^N \times \mathbf{R} \times \mathbf{R}^N$ and $B: Q \to \mathbf{R}$ is scalar, subject to two kinds of restrictions. The first is the *ellipticity condition*, an essential requirement in the theory. It consists in asking the matrix $\{a_{ij} = \partial A_i/\partial \xi_j\}$ to be definite positive, strictly or not, uniformly or not, with respect to x. The second usual requirement are the *structure conditions* which refer to the growth allowed to the nonlinearities $\mathcal{A}$ and B as well to their integrability properties. A typical requirement is that

(0.2) $$\begin{aligned} \mathcal{A}(x, u, \xi)\xi &\geq c_1(x)|\xi|^p - c_2(x)|u|^p - c_3(x), \\ |\mathcal{A}(x, u, \xi)| &\leq c_4(x)|\xi|^{p-1} + c_5(x)|u|^{p-1} + c_6(x), \\ |B(x, u, \xi)| &\leq c_7(x)|\xi|^{p-1} + c_8(x)|u|^{p-1} + c_9(x), \end{aligned}$$

for some number $p > 1$ and suitable functions $c_i(x)$, with $c_1(x) > 0$, cf. [S1]. This type of equations has been intensively studied in the last half century and a huge number of results have been proved for quite general classes of such equations, cf. e.g., [LU], [GT]. Many of those results preserve a flavor of the most basic example, the Laplace-Poisson equation $\Delta u = f$, while others depart strongly from it. Thus, one of the most celebrated results in nonlinear PDE's in the century has been the proof by De Giorgi and Moser of the property of Hölder continuity for the solutions of equations of the form $\sum \partial_i(a_{ij}\partial_j u) = 0$ with just bounded measurable coefficients forming a uniformly elliptic matrix. This result has been generalized subsequently to quite general equations of the above form, on the assumption that $\mathcal{A}$ and B are well-behaved. The question of $C^{1,\alpha}$ regularity was studied in the classical works of Ladyzhenskaya and Uraltseva, Morrey, Uhlenbeck, Evans, ... and the study continues in different directions.

There is however another direction of research where the functions $\mathcal{A}$ or B include unbounded, merely integrable terms, and then we can not expect the solutions to be smooth, even continuous. Problems of existence and uniquenss naturally ensue. Successful theories have been based on energy methods, so-called variational approach, and can be seen developed in some detail in texts like the ones already mentioned.

Here I will present a problem of current interest in the existence and uniqueness theory. It concerns the definition of a well-posed problem for general types of quasi-linear equations when the data are taken in classes of integrable functions (or even Borel measures) so that the variational approach does not apply. Already in 1964 J. Serrin pointed out in [S2] that distribution solutions are in general not unique by means of an explicit example for the linear equation

$$(0.3) \qquad \sum \frac{\partial}{\partial x_i}\left(a_{ij}(x)\frac{\partial u}{\partial x_j}\right) = f(x)$$

in a ball B, with a bad but bounded and uniformly elliptic matrix $\{a_{ij}\}$. The 'bad' solution has a point singularity and belongs to $W^{1,1}(B)$ but not to the energy space $W^{1,2}(B)$. See discussion in [B6].

Substantial progress in the problem of characterizing a class of uniqueness has been achieved recently by several authors. In this article, which is an expanded version of the lecture given at the Fez Conference, I will explain the concept of *entropy solution*, as developed by Ph. Bénilan, L. Boccardo, Th. Gallouët, R. Gariepy, M. Pierre and the author in [B6]. Our main results can be summarized as follows:

We define a concept of entropy solution for the above type of equations by means a family of inequalities in terms of truncated functions. This concept makes sense for all growth exponents $p > 1$. Under standard structure conditions we obtain existence and uniqueness of an entropy solution for every $f \in L^1(\Omega)$.

I will comment in some detail the main issues, using for clarity the so-called p-Laplacian example. I will also give in detail the uniqueness proof of [B6], since it is particularly simple, to which I will add continuous dependence estimates. There is also for the record a derivation of the a main priori estimates at variance from the one we finally included in [B6].

I will also comment on parallel work of P. L. Lions and F. Murat on the concept of *renormalized solutions*, introduced years ago by R. Di Perna and P. L. Lions [DL] in the study of the Botzmann equations. The equivalence of both concepts for the simplest models is shown. The proof depends on two facts: (i) for good data we get the usual variational solutions, which are at the same time entropy and renormalized ones; (ii) the new types of solutions exist and are unique for all data $f \in L^1(\Omega)$ and depend continuously in suitable norms on the data. Finally, I will briefly comment on the works of several other authors who have recently contributed interesting results to this area of research. But I will center on matters related to the entropy solutions with no pretense of a comprehensive discussion of the literature, and I apologize for any omissions. The reader will be interested in reading L. Boccardo's contribution in this volume where convergence issues are carefully discussed.

1. Basic problem. Entropy solutions

The problem can be discussed most clearly if we consider the model equation

(1.1) $$-\text{div}\,(|Du|^{p-2}Du) + \lambda u = f(x) \quad \text{in } \Omega,$$

where $1 < p < \infty$, Ω is an open subset of $\mathbf{R}^N$ ($N \geq 1$), not necessarily bounded, $Du = (\partial_1 u, \cdots, \partial_N u)$ denotes the gradient of u and $\lambda \geq 0$ is a constant. We will use the p-Laplacian notation, $\Delta_p(u)$, for the operator $\text{div}\,(|Du|^{p-2}Du)$. This equation reflects the main difficulties we will have to tackle. Later in Section 8 we will make a general statement of the equations we can cover with the same techniques. Let us also remark that the term λu plays for $\lambda > 0$ a regularizing role that makes some of the analysis easier, so that the main difficulties will concern the case $\lambda = 0$. For definiteness we will consider the homogeneous Dirichlet problem

(1.2) $$u(x) = 0 \quad \text{for } x \in \partial\Omega.$$

When Ω is bounded or $\lambda > 0$ the standard variational theory asserts that for every $f \in W^{-1,p'}(\Omega)$, $p' = p/(p-1)$, there exists a unique $u \in W_0^{1,p}(\Omega)$ which is a weak solution of problem (1.1)-(1.2), i.e.,

(1.3) $$\int_\Omega |Du|^{p-2} \langle Du, D\phi\rangle\, dx + \lambda \int_\Omega u\phi\, dx = \int_\Omega f\phi\, dx$$

for every $\phi \in W_0^{1,p}(\Omega)$. However, our emphasis is placed on solving this equation under the requirement that $f(x) \in L^1(\Omega)$. For the range of exponents $p > N$ there is no novelty since $L^1(\Omega) \subset W^{-1,p'}(\Omega)$. In other words, one can get estimates for the solutions of the variational problem which prove that u is bounded in terms of $\|f\|_1$ and then the gradient, Du, belongs to $L^p(\Omega)$ and (1.3) holds, cf. [LL]. Also, when $p = 2$ we have the classical linear equation $\Delta u = f$ and we can apply the Green function technique to get a well-posed problem.

The situation is different for $p \leq N$, $p \neq 2$, and we have to use a a new framework for our theory. Indeed, there are two difficulties associated with the study of equation (1.1), even in a bounded domain. Let us also point out that linear techniques cannot be applied to nonlinear equations of the form (0.1), (0.2) even if they have linear growth ($p = 2$) and our results have an interest also in that case.

We propose to solve the problem by approximation with variational solutions: we take $f_n \in C_0(\Omega)$ such that $f_n \to f$ in $L^1(\Omega)$, we find a solution $u_n \in W_0^{1,p}(\Omega)$ for the problem with second member f_n and we will try to pass to the limit $n \to \infty$. In doing so we find that the natural estimates for u and Du are the following. Let us point out that we are interested in estimates which do not depend on λ.

Lemma 1.1. *Let $1 < p < N$ and let u be a variational solution of (1.1)-(1.2) with $f \in L^1(\Omega)$. Then there exists $C = C(N,p) > 0$ such that*

(1.4) $$\text{meas}\,\{|u| > k\} \leq C\,M^{\frac{N}{N-p}}\,k^{-p_1},$$

where $M = \|f\|_1$ and $p_1 = \frac{N(p-1)}{N-p}$.

In functional terms this means that $u \in \mathcal{M}^{p_1}(\Omega)$. We recall, cf. [BBC], that for $1 < q < \infty$ the Marcinkiewicz space $\mathcal{M}^q(\Omega)$ can be defined as the set of measurable functions $u : \Omega \to \mathbf{R}$ such that the corresponding distribution functions

(1.5) $$\Phi_u(k) = \text{meas}\,\{x \in \Omega : |u(x)| > k\}$$

satisfy an estimate of the form

(1.7) $$\Phi_u(k) \leq C\,k^{-q}, \qquad C < \infty.$$

It is immediate that $L^q(\Omega) \subset \mathcal{M}^q(\Omega)$ and that for bounded Ω we have $\mathcal{M}^q(\Omega) \subset \mathcal{M}^{\hat{q}}(\Omega)$ if $q \geq \hat{q}$.

Lemma 1.2. *Let $1 < p < N$ and let u be a solution as above. Then for every $h > 0$*

(1.8) $$\text{meas}\,\{|Du| > h\} \leq C(N,p)\,M^{\frac{N}{N-1}}\,h^{-p_2}, \qquad p_2 = \frac{N(p-1)}{N-1}.$$

Hence, $|Du| \in \mathcal{M}^{p_2}(\Omega)$.

These estimates are optimal in the present setting as can be checked by inspection of the fundamental solution, i.e. the "solution" of (1.1) when $f = c\delta(0)$, a Dirac mass, which in a ball $\Omega = B_R$ takes the form

(1.9) $$U(x) = C(|x|^{-\alpha} - R^{-\alpha}), \qquad \alpha = \frac{N-p}{p-1}.$$

This is a limit situation for the L^1 theory. We see that U and $|DU|$ belong precisely to the above-mentioned spaces. The main conclusion we draw from these estimates and example is that we face a serious **difficulty**: we do not get the variational property $|Du| \in L^p(\Omega)$ and this will spoil the standard uniqueness theory. To make things worse we only get $|DU| \in L^1(B_R)$ if $p > p_0$ with

(1.10) $$p_0 = 2 - (1/N),$$

so that we may expect big trouble for the range of very small p's, $1 < p < p_0$, which exists for $N \geq 2$. However, not everything is lost. At least we get $|DU|^{p-2}DU \in L^1(\Omega)$ so the term that actually enters the equation is controlled. On the other hand, we get a hint of the way towards a solution of the difficulties in the form of an estimate for truncated functions, $T_k(u)$ defined by

(1.11) $$T_k(u) = \begin{cases} u & \text{if } |s| \leq k \\ k\,\text{sign}\,(s) & \text{if } |s| > k. \end{cases}$$

We will state a third basic estimate

Lemma 1.3. *Let u be a variational solution of (1.1)-(1.2) with $f \in L^1(\Omega)$. Then for every $k > 0$*

(1.12) $$\frac{1}{k}\|DT_k(u)\|_p^p = \frac{1}{k}\int_{\{|u|<k\}} |Du|^p dx \leq \int_{\{|u|<k\}} |f|\, dx \leq \|f\|_1.$$

Moreover, for k and $h > 0$

(1.13) $$\frac{1}{h}\int_{\{k<|u|<k+h\}} |Du|^p dx \leq \int_{\{k<|u|<k+h\}} |f|\, dx \leq \|f\|_1.$$

Observe that the intermediate term goes to zero as $k \to \infty$ since f is integrable. On the other hand, we see that that even if we cannot control $|Du|$ globally in Ω we can control certain combinations of $|Du|$ and u. Indeed, for all $\rho > 0$ we have

(1.14) $$\int_{\{|u|>k\}} \frac{|Du|^p}{(1+|u|)^{1+\rho}}\, dx \leq \frac{C}{\rho(1+k)^\rho}\|f\|_1.$$

Let us now state the main consequence of making the assumption $\lambda > 0$ in (1.1).

Lemma 1.4. *Under the above assumptions let $\lambda > 0$. Then $u \in L^1(\Omega)$ and*

(1.15) $$\lambda \int_\Omega |u|\, dx \leq \int_\Omega |f|\, dx.$$

With these estimates we can propose a method based on approximation by regular problems and passing to the limit in the approximations obtained. We will assume in a first step that $p > p_0$. We state next the basic existence theorem for problem (1.1)-(1.2).

Theorem 1.5. *Under the assumption $p_0 < p < N$ there exists a function*

(1.16) $$u \in M^{p_1}(\Omega), \quad |Du| \in M^{p_2}(\Omega)$$

where $p_1 = \frac{N(p-1)}{N-p}$ and $p_2 = \frac{N(p-1)}{N-1}$, which is a solution of (1.1) in the sense of distributions. It follows that $u \in W^{1,q}_{loc}(\Omega)$ for every $q < p_2$. If Ω is bounded then $u \in W^{1,q}_0(\Omega)$. Otherwise, the boundary condition (1.2) is taken in the sense that for every cutoff function $\zeta \in C_0^\infty(\mathbf{R}^N)$ we have $\zeta T_k(u) \in W^{1,p}_0(\Omega)$. If $\lambda > 0$ then $u \in L^1(\Omega)$.

As we advanced, a difficulty appears with the question of uniqueness of such solutions. Actually, we cannot obtain uniqueness of distribution solutions of (1.1)-(1.2) in the framework of Theorem 1.5. We obtain uniqueness of a special class of distribution solutions that satisfy an extra condition that we call the *entropy condition*.

Definition 1.6. A measurable function $u : \Omega \to \mathbf{R}$ such that $\zeta T_k(u) \in W^{1,p}_{loc}(\Omega)$ is called an **entropy solution** of problem (1.1)-(1.2) if it satisfies the family of inequalities

$$(1.17) \qquad \int_{\{|u-\phi|<k\}} |Du|^{p-2} \langle Du, Du - D\phi \rangle \, dx \leq \int_{\Omega} T_k(u - \phi)(f - \lambda u) \, dx,$$

for every $\phi \in \mathcal{D}(\Omega)$ and every $k > 0$. ζ is a standard cutoff function in $\mathbf{R}^N$ and T_k is defined in (1.11).

The use of such conditions is rather common in hyperbolic conservation laws after the work of Kruzhkov, [Kr], but is novel to elliptic equations. Our main result is the following.

Theorem 1.7. *The entropy solution of problem (1.1), (1.2) in the framework of Theorem 1.5 is unique unless $\Omega = \mathbf{R}^N$ and $p \geq N$ in which case it is unique up to constants.*

2. The problem for small p

We now turn our attention to the **second difficulty** of the problem, namely how to give sense to the solutions corresponding to small exponents p in the range $1 < p \leq p_0$, $N \geq 2$. The preceding discussion pointed out that for general $f \in L^1(\Omega)$ we cannot expect the solution to be in $W^{1,1}_{loc}(\Omega)$ in this range of p, i.e., we cannot take the gradient of u appearing in the p-Laplacian operator in the usual distribution sense. So there is a problem of how to give sense to the solutions of equation (1.1). We solve this difficulty in [B6] by elaborating on the idea of the truncated functions, which allows to define a new space $\mathcal{T}^{1,1}_{loc}(\Omega)$ in which we can naturally give a sense to the gradient of u that generally is *not locally integrable*. Here are the relevant definitions.

Definition 2.1. i) The space $\mathcal{T}^{1,1}_{loc}(\Omega)$ is defined as the set of measurable functions $u : \Omega \mapsto \mathbf{R}$ such that for every $k > 0$ the truncated function $T_k(u)$ belongs to $W^{1,1}_{loc}(\Omega)$.

ii) The space $\mathcal{T}^{1,p}_{loc}(\Omega)$, $p \in (1, \infty)$, is the subset of $\mathcal{T}^{1,1}_{loc}(\Omega)$ consisting of the functions u such that $D(T_k(u)) \in L^p_{loc}(\Omega)$ for every $k > 0$. Likewise, $\mathcal{T}^{1,p}(\Omega)$ is the subset of $\mathcal{T}^{1,1}_{loc}(\Omega)$ consisting of the u such that moreover $DT_k(u) \in L^p(\Omega)$ for every $k > 0$.

iii) Finally, $\mathcal{T}^{1,p}_0(\Omega)$ will be the subset of $\mathcal{T}^{1,p}(\Omega)$ consisting of the functions that can be approximated by smooth functions with compact support in Ω in the following sense: a function $u \in \mathcal{T}^{1,p}(\Omega)$ belongs to $\mathcal{T}^{1,p}_0(\Omega)$ if for every $k > 0$

there exists a sequence $\phi_n \in C_0^\infty(\Omega)$ such that

$$D\phi_n \to DT_k(u) \quad \text{in} \quad L^p(\Omega),$$
$$\phi_n \to T_k(u) \quad \text{in} \quad L^1_{loc}(\Omega).$$

There are a number of alternative characterizations of $T_0^{1,p}(\Omega)$ that can be consulted in [B6]. A very useful one is

Lemma 2.2. *$u \in T_0^{1,p}(\Omega)$ if and only if $u \in T^{1,p}(\Omega)$ and for every $k > 0$ and every smooth cutoff function $\zeta \in C_0^\infty(\mathbf{R}^N)$*
  (a) $\zeta T_k(u) \in W_0^{1,p}(\Omega)$,
  (b) if $p < N$ and Ω is not bounded we also need the condition $T_k(u) \to 0$ in measure as $|x| \to \infty$. In other words, $u \in L_0(\Omega)$.

$L_0(\Omega)$ is the set of measurable functions tending to 0 in measure as x goes to ∞, cf. [B6]. For every $p \in [1,\infty)$ we have the inclusions $W_{loc}^{1,p}(\Omega) \subset T_{loc}^{1,p}(\Omega)$ and $W_0^{1,p}(\Omega) \subset T_0^{1,p}(\Omega)$ and in these cases we have

(2.1) $$DT_k(u) = 1_{\{|u|<k\}} Du,$$

where 1_A denotes the characteristic function of a measurable set $A \subset \mathbf{R}^N$. It is also clear that $T_{loc}^{1,p}(\Omega) \cap L_{loc}^\infty(\Omega) = W_{loc}^{1,p}(\Omega) \cap L_{loc}^\infty(\Omega)$. Moreover, we can easily convince ourselves that the inclusions are strict, i.e. the new spaces are strict extensions. But we have to mention that $T_{loc}^{1,1}(\Omega)$ is not even a vector space. Let us also mention that we did not impose the condition $T_k(u) \in L^p(\Omega)$. Of course, this condition follows immediately when Ω has finite measure (since $T_k(u)$ is bounded), but for unbounded Ω it makes a real difference. Of course, we have: if $u \in T_0^{1,p}(\Omega)$ and $1 < p < N$ then $DT_k(u) \in L^p(\Omega)$ and $T_k(u) \in L^{p^*}(\Omega)$ for $p^* = pN/(N-p)$. This in particular implies that for $1 < p < N$, $u \to 0$ in measure as $|x| \to \infty$, which is not necessarily the case for $p \geq N$. The simpler situation for bounded domains is summarized as follows:

Lemma 2.3. *Let Ω be bounded and $1 < p < \infty$. Then $u \in T_0^{1,p}(\Omega)$ iff $T_k(u) \in W_0^{1,p}(\Omega)$ for every $k > 0$ and $u \in W_0^{1,p}(\Omega)$ iff $u \in T_0^{1,p}(\Omega)$ and $Du \in L^p(\Omega)$.*

The main point of these functional setting is that it allows to give a sense to the derivative Du of a function $u \in T_{loc}^{1,1}(\Omega)$, generalizing the usual concept of weak derivative in $W_{loc}^{1,1}(\Omega)$, cf. [GT].

Lemma 2.4. *For every $u \in T_{loc}^{1,1}(\Omega)$ there exists a unique measurable function $v : \Omega \mapsto \mathbf{R}^N$ such that*

(2.2) $$DT_k(u) = v 1_{\{|v|<k\}} \quad \text{a.e.}$$

The function v is called the **generalized derivative** of u. We have $u \in W^{1,1}_{loc}(\Omega)$ if and only if $v \in L^1_{loc}(\Omega)$, and then $v \equiv Du$ in the usual weak sense.

In this framework the first term in equation (1.1) makes sense when $|Du|^{p-2}Du \in L^1_{loc}(\Omega)$. In order to take into account condition (1.2) we seek the solution in $T^{1,p}_0(\Omega)$.

The definition of entropy solution (1.17) makes perfect sense for functions in $T^{1,p}_0(\Omega)$. It is immediate that a variational solution (where $Du \in L^p(\Omega)$) is an entropy solution and conversely that an entropy solution with $|Du| \in L^p(\Omega)$ is a variational solution. It is also proved in [B6] that entropy solutions in $T^{1,p}_0(\Omega)$ are distributional solutions. With these definitions we can extend the existence and uniqueness results to the whole interval $1 < p < N$. Theorem 1.5 is extended to the whole range $(1, N)$ like this.

Theorem 2.5. *For every $1 < p < N$ there exists a function $u \in T^{1,p}_0(\Omega)$ such that*

$$u \in \mathcal{M}^{p_1}(\Omega), \quad |Du| \in \mathcal{M}^{p_2}(\Omega),$$

where $p_1 = N(p-1)/(N-p)$ and $p_2 = N(p-1)/(N-1)$, which is a solution of (1.1) in the sense of distributions and in the entropy sense. The boundary condition is contained in the definition of $T^{1,p}_0(\Omega)$.

The uniqueness result, Theorem 1.7, admits the extension, valid for all $p > 1$.

Theorem 2.6. *The entropy solution of (1.1), (1.2) in the class $T^{1,p}_0(\Omega)$ is unique with the proviso of Theorem 1.7.*

The proof of Theorem 2.5 is sketched in Section 4 and that of Theorem 2.6 in Section 5. Let me point out that some of the basic ideas of this developement, including the introduction of T-spaces to account for the unusual derivatives and the a priori estimates of the distribution function of u and Du stem from the late 80's, cf. [Be]. The formulation of the entropy condition and the corresponding uniqueness is the most novel part of the work, contained in [B6].

3. The a priori estimates

Let us prove the a priori estimates for variational solutions announced in Lemmas 1.1 and 1.2.

PROOF OF LEMMAS 1.3 AND 1.4. To get (1.12) we multiply the equation by $T_k(u)$ and integrate by parts to get

$$\int_{\{|u|<k\}} |Du|^p \, dx + \int_\Omega \lambda u \, T_k(u) \, dx = \int_\Omega f T_k(u) \, dx.$$

For (1.15) we divide by k and take the limit as $k \to 0$. For (1.13) the multiplier is $T_{k,h}(u)$, where
$$T_{k,h}(s) = \min\{\max\{s-k, 0\}, h\}.$$
From this it follows that
$$\int_{\{|u|\geq k\}} \frac{|Du|^p}{(1+|u|)^{1+\rho}} dx \leq \int_k^\infty \frac{dh}{(1+h)^{1+\rho}} \|f\|_1.$$

We see that this is bounded above by $c(1+k)^{-\rho}$ with $c = \|f\|_1/\rho$, hence (1.14).

PROOF OF LEMMA 1.1. Let $\Phi(k)$ be the distribution function of u and let
$$F_\varepsilon(r) = \min\{\frac{r^+}{\varepsilon}, 1\} = \frac{1}{\varepsilon}(T_\varepsilon(r))_+,$$
so that $F_\varepsilon(r)$ is nondecreasing, $F_\varepsilon(r) = r/\varepsilon$ for $0 < r < \varepsilon$, $F_\varepsilon(r) = 0$ for $r \leq 0$ and $F_\varepsilon(r) = 1$ for $r \geq \varepsilon$. We apply the usual Sobolev inequality to the function $F_\varepsilon(|u| - k)$ with k and $\varepsilon > 0$ to get

(3.1) $$\|F_\varepsilon(|u| - k)\|_{L^{N/(N-1)}} \leq c_N \int_{\{k<|u|<k+\varepsilon\}} \frac{|Du|}{\varepsilon} dx,$$

where c_N is the Sobolev constant of the embedding $W^{1,1}(\mathbf{R}^N) \to L^{N/(N-1)}(\mathbf{R}^N)$. It is also clear that for every $s > 0$

(3.2) $$\Phi(k+\varepsilon) \leq \int (F_\varepsilon(|u| - k))^s dx.$$

On the other hand, we have
$$\int_{\{k<|u|<k+\varepsilon\}} |Du| dx \leq$$
$$\leq (\int_{\{k<|u|<k+\varepsilon\}} |Du|^p dx)^{1/p} \text{meas}\{k < |u| < k+\varepsilon\}^{(p-1)/p}.$$

Combining this with Lemma 1.3, (3.1) and (3.2) with $s = N/(N-1)$ we get

(3.3) $$\Phi(k+\varepsilon)^\alpha \leq C_1 \frac{\Phi(k) - \Phi(k+\varepsilon)}{\varepsilon}, \quad \text{with} \quad \alpha = \frac{(N-1)p}{N(p-1)}.$$

The constant is $C_1 = c_N^{p/(p-1)} M^{1/(p-1)}$. Let now $\varepsilon \to 0$. Then, (3.3) becomes

(3.4) $$(-\Phi'(k)) \geq \frac{1}{C_1} \Phi(k)^\alpha.$$

[The derivative of Φ is the generalized derivative of a monotone function, in principle a measure]. We obtain the result after integration of this differential inequality. It is at this stage that the relative value of p and N plays a role. Since $p < N$ the exponent α is larger than 1. Actually, $\alpha = 1 + (1/p_1)$. Then (3.4) implies that

$$(\Phi^{-\frac{1}{p_1}})' \geq \frac{1}{p_1 C_1},$$

and we get (1.4) by integration. The resulting constant is precisely

$$C(N,p) = (p_1\, c_N^{\frac{p}{p-1}})^{p_1}.$$

PROOF OF LEMMA 1.2. Let $\theta \in (0,1)$ and $k > 0$. For any set $A \subset \Omega$

(3.5) $$\int_A |Du|^{\theta p}\, dx \leq \int_{\{|u|\geq k\}} |Du|^{\theta p}\, dx + \int_{\{|u|<k\}\cap A} |Du|^{\theta p}\, dx.$$

The last integral can be estimated as

$$\int_{\{|u|<k\}\cap A} |Du|^{\theta p}\, dx \leq \left(\int_{\{|u|<k\}\cap A} |Du|^p\, dx\right)^{\theta} (\mathrm{meas}(\{|u|<k\}\cap A))^{1-\theta}$$

$$\leq |A|^{1-\theta}\left(\int_{\{|u|<k\}} |Du|^p\, dx\right)^{\theta} \leq |A|^{1-\theta}(k\,\|f\|_1)^{\theta},$$

where $|A| = \mathrm{meas}(A)$. The last inequality is a consequence of the entropy definition. On the other hand, given a $\rho > 0$ we may estimate the other integral in (3.5) as

(3.6)
$$\int_{\{|u|\geq k\}} |Du|^{\theta p}\, dx = \int_{\{|u|\geq k\}} |u|^{(1+\rho)\theta}\left(\frac{|Du|^p}{|u|^{1+\rho}}\right)^{\theta} dx$$

$$\leq \left[\int_{\{|u|\geq k\}} \frac{|Du|^p}{|u|^{1+\rho}}\, dx\right]^{\theta}\left[\int_{\{|u|\geq k\}} |u|^{\frac{(1+\rho)\theta}{1-\theta}}\, dx\right]^{1-\theta} = I_1^{\theta}\cdot I_2^{1-\theta}.$$

We have to estimate I_1 and I_2. In order to estimate I_1 we use (1.14) (a consequence of Lemma 1.3) to see that it is bounded above by $c\,k^{-\rho}$ with $c = M/\rho$, $M = \|f\|_1$. In order to estimate the other integral, I_2, we put

(3.7) $$p_1 = \frac{(p-1)N}{N-p} \quad \text{and} \quad \gamma = \frac{(1+\rho)\theta}{1-\theta}.$$

We can always choose ρ and θ so small that $\gamma < p_1$. Then, with Φ the distribution function of u and using (1.4) we get

$$\int_{\{|u|\geq k\}} |u|^\gamma \, dx = -\int_k^\infty t^\gamma \, d\Phi(t) = k^\gamma \Phi(k) + \int_k^\infty \gamma t^{\gamma-1} \Phi(t) \, dt$$

$$\leq k^\gamma \frac{C_1}{k^{p_1}} + \gamma C_1 \int_k^\infty t^{\gamma-p_1-1} \, dt = \frac{C_1 p_1}{(p_1-\gamma) k^{p_1-\gamma}}.$$

($C_1 = CM^{\frac{N}{N-p}}$ is the constant of (1.4)). Putting together all the pieces, (3.5) becomes

(3.8)
$$\int_A |Du|^{\theta p} \, dx \leq |A|^{1-\theta} (Mk)^\theta + \frac{(M/\rho)^\theta}{k^{\rho\theta}} \frac{(C_1 p_1)^{1-\theta}}{(p_1-\gamma)^{1-\theta} k^{(p_1-\gamma)(1-\theta)}}$$

$$\leq c_1 (|A|^{1-\theta} k^\theta + \frac{c_2}{k^{\mu\theta}})$$

where $\mu = \rho + (p_1 - \gamma)(1-\theta)/\theta$ and the constants c_1 and c_2 depend on M in some power way. We minimize the second member of this inequality with respect to the variable k. The minimum is obtained for

$$k = c_3 |A|^{\frac{\theta-1}{\theta(\mu+1)}}.$$

With this value (3.8) becomes

$$\int_A |Du|^{\theta p} \, dx \leq c_4 |A|^{\frac{(1-\theta)\mu}{\mu+1}}.$$

Take now $A = \{|Du| > \lambda\}$. We get

$$\lambda^{\theta p} |A| \leq \int_A |Du|^{\theta p} \, dx \leq c_4 |A|^{\frac{(1-\theta)\mu}{\mu+1}},$$

so that

$$|A|^{1-\frac{(1-\theta)\mu}{\mu+1}} \leq c_4 \lambda^{-\theta p},$$

namely,

$$|A| \leq c_5 \lambda^{-\frac{\theta p(\mu+1)}{1+\theta\mu}}.$$

Though you might find it fantastic, the last exponent is just $N(p-1)/(N-1)$. Moreover, following more closely the dependence of the constants c_i gives the dependence stated in Lemma 1.2 (we can be sure of this form of dependence since it can be obtained more directly by dimensional considerations, i.e. scaling). The proof is finished.

Let us point out that there are a number of alternative proofs of these facts, among them the ones given in [B6]. Better regularity can be obtained under more restrictive assumptions on f, cf. [BG1], [BG2]. Thus, in the latter it was proved that we can obtain the regularity $u \in W_0^{1,p_2}(\Omega)$ provided that

(3.9) $\qquad f \in L^1 \log L^1(\Omega)$, i.e. $\displaystyle\int_\Omega |f|\log(1+|f|)dx < \infty.$

4. Proof of existence

Our FIRST STEP consists in approximating the second member f with a sequence of smooth functions $f_n \in C_0^\infty(\Omega)$, $f_n \to f$ in $L^1(\Omega)$. It will be also useful to ask that

(4.1) $\qquad \|f_n\|_1 \leq \|f\|_1$

for every $n \geq 1$. If $\lambda = 0$ we add a regularization $\lambda_n(s) = (1/n)|s|^{p-2}s$. Otherwise we put $\lambda_n(s) = \lambda(s)$. Then it is well-known, see [LL], [Li], that for any domain there exists a unique $u_n \in W_0^{1,p}(\Omega)$ such that

(4.2) $\qquad -\Delta_p(u_n) + \lambda_n(u_n) = f_n$

holds in the sense of distributions in Ω. We also point out that $u_n \in L^1(\Omega) \cap L^\infty(\Omega)$. Multiplying (4.2) by convenient test functions and integrating one gets the following uniform estimates

(4.3) $\qquad \dfrac{1}{a}\displaystyle\int_{\{k<|u_n|<k+a\}} |Du_n|^p\,dx \leq \int_{\{|u_n|>k\}} |f_n|\,dx \leq \|f_n\|_1 = C_1.$

(4.4) $\qquad \displaystyle\int_{\{|u_n|>k\}} |\lambda_n(u_n)|\,dx \leq \int_{\{|u_n|>k\}} |f_n|\,dx \leq \|f_n\|_1 \leq C_1.$

(4.5) $\qquad \displaystyle\int_{\{|u_n|<k\}} |Du_n|^p\,dx \leq kC_1.$

In a SECOND STEP and based on these estimates we can show that $\{T_k u_n\}_n$ is relatively compact in $L^{p^*}(\Omega)$ and that

$$\operatorname{meas}(\{|T_k u_n - T_k u_m| > t\} \cap B_R) \leq t^{-q}\int_{\Omega \cap B_R} |T_k u_n - T_k u_m|^q\,dx \leq \varepsilon$$

for all $n,m \geq n_0(k,t,R)$. This proves that $\{u_n\}$ is a *Cauchy sequence in measure* in $\Omega \cap B_R$, hence that $u_n \to u$ locally in measure.

We then prove that Du_n converges to some function v *locally in measure* (and therefore, we can always assume that the convergence is a.e. after passing to a suitable subsequence). Moreover, since $\{DT_k u_n\}_n$ is bounded in $L^p(\Omega)$ (for any $k > 0$), it converges weakly to $D(T_k u)$ in $L^1_{loc}(\Omega)$. Then, we have $u \in T^{1,1}_{loc}(\Omega)$ and $Du = v$ a.e.

We also have $D(T_k(u)) \in L^p(\Omega)$ and moreover $u \in T_0^{1,p}(\Omega)$. Indeed, we can construct $\phi_n \in C_0^\infty(\Omega)$ such that $\|D\phi_n - D(T_k u_n)\|_{L^p} \leq 1/n$ and $\|\phi_n - T_k u_n\|_{L^{p_*}} \leq 1/n$. We then have $D\phi_n \to D(T_k u)$ weakly in $L^p(\Omega)$ and $\phi_n \to T_k u$ strongly in $L^q_{loc}(\Omega)$ for $q < p_*$. From ϕ_n we can construct ψ_n (convex combinations of the ϕ_n's, using Mazur's lemma) so as to have strong convergence of derivatives. We conclude that $u \in T_0^{1,p}(\Omega)$.

Furthermore, using the convergence of u_n to u and Du_n to Du, we can prove for u the a priori estimates of Lemmas 1.1 and 1.2.

The NEXT STEP is proving that $|Du|^{p-2} Du \in L^1_{loc}(\Omega)$ and

$$|Du_n|^{p-2} Du_n \to |Du|^{p-2} Du \quad \text{strongly in} \quad L^1_{loc}(\Omega).$$

It then immediately follows that

$$-\text{div}\,(|Du|^{p-2} Du)) + \lambda u = f \quad \text{in } \mathcal{D}'(\Omega).$$

The FINAL STEP consists in checking that the entropy inequality holds. Actually, it is possible to prove that equality holds in the entropy formulas by proving that for all $k > 0$

$$D(T(u_n)) \to D(T(u)) \quad \text{in} \quad L^p(\Omega).$$

Complete details can be found in [B6].

5. Uniqueness of entropy solutions

The proof is based on repeating the idea of the uniqueness proof in the variational framework with proper changes. The main obstacle is that we cannot use the difference of two solutions, $u_1 - u_2$, as a test function. However, the entropy definition provides us with a set of variational-like inequalities which we can use to the effect. The key tecnical point in [B6] is that certain combinations of truncations of the solutions will do the job. Before we start we need the following technical result that increases our store of admissible test functions.

Lemma 5.1. *If u is an entropy solution of (1.1)-(1.2). Then (1.17) holds for every test function $\phi \in T_0^{1,p}(\Omega) \cap L^\infty(\Omega)$.*

We now proceed to sketch the proof given in [B6], simplified to our more particular problem. Let u_i, $i = 1, 2$, be two solutions with second member $f(x) \in L^1(\Omega)$. Using Lemma 5.1 we write the entropy inequality corresponding to solution u_1 with test function $T_h u_2$ and u_2 with test function $T_h u_1$. Adding up both results we get

(5.1)
$$\int_{\{|u_1-T_h u_2|<k\}} |Du_1|^{p-2}\langle Du_1, Du_1 - DT_h u_2\rangle dx+$$
$$\int_{\{|u_2-T_h u_1|<k\}} |Du_2|^{p-2}\langle Du_2, Du_2 - DT_h u_1\rangle dx+$$
$$\int_\Omega \lambda\left[u_1 T_k(u_1 - T_h(u_2)) + u_2 T_k(u_2 - T_h(u_1))\right] dx \leq$$
$$\int_\Omega f\left[T_k(u_1 - T_h(u_2)) + T_k(u_2 - T_h(u_1))\right] dx.$$

(ii) We split the integrals in the first two terms of (5.1) into the contributions corresponding to different integration sets. Then the main contribution of the first member of (5.1) is located in the set

$$A_0 = \{x \in \Omega : |u_1 - u_2| < k, \ |u_1| < h, \ |u_2| < h\}$$

and amounts to

$$I_0 = \int_{A_0} \langle |Du_1|^{p-2} Du_1 - |Du_2|^{p-2} Du_2, Du_1 - Du_2\rangle dx.$$

On the other hand, on the set $A_1 = \{x \in \Omega : |u_1 - T_h u_2| < k, |u_2| \geq h\}$ we have

$$\int_{A_1} \langle |Du_1|^{p-2} Du_1, Du_1 - DT_h u_2\rangle dx = \int_{A_1} \langle |Du_1|^{p-2} Du_1, Du_1\rangle dx \geq 0,$$

while on the remaining set $A_2 = \{x \in \Omega : |u_1 - T_h u_2| < k, |u_2| < h, |u_1| \geq h\}$ we get

$$\int_{A_2} \langle |Du_1|^{p-2} Du_1, Du_1 - DT_h u_2\rangle dx = \int_{A_2} \langle |Du_1|^{p-2} Du_1, Du_1 - Du_2\rangle dx$$
$$\geq - \int_{A_2} \langle |Du_1|^{p-2} Du_1, Du_2\rangle dx.$$

In the same way we estimate the second integral in the sets A_1', where $|u_1| \geq h$, and A_2', where $|u_1| < h$ and $|u_2| \geq h$. All these sets and integrals depend of course on k and h. Summing up we estimate the first member of (5.1) in the form $I \geq I_0 - I_1$, where

$$I_1 = \int_{A_2} \langle |Du_1|^{p-2} Du_1, Du_2\rangle dx + \int_{A_2'} \langle |Du_2|^{p-2} Du_1, Du_1\rangle dx.$$

Now, it is easily proved using Lemma 1.3 that I_1 goes to 0 as $h \to \infty$ [since $\int_K f dx \to 0$ as $\text{meas}(K) \to 0$].

(iii) The second member of (5.1) can be worked out more easily. On the set $B_0 = \{x \in \Omega : |u_1| < h, |u_2| < h\}$ the integrand is zero, while on the sets $B_1 = \{x \in \Omega : |u_1| \geq h\}$ and $B_2 = \{x \in \Omega : |u_2| \geq h\}$ the integrals can be estimated by

$$|J_i| \leq 2k \int_{B_i} (|f|)\, dx.$$

Now, the measure of both sets, $B_1(h,k)$ and $B_2(h,k)$, goes to zero as $h \to \infty$ for fixed $k > 0$. Hence $J_1 + J_2 \to 0$.

(iv) The terms in λ give in a similar way as $h \to \infty$ the contribution

(5.2) $$\lambda \int_\Omega (u_1 - u_2) T_k(u_1 - u_2)\, dx \geq 0$$

(v) Combining the above estimates we get from (5.1)

$$\int_{A_0(h,k)} \langle |Du_1|^{p-2} Du_1 - |Du_2|^{p-2} Du_2, Du_1 - Du_2 \rangle dx \leq \varepsilon(h),$$

where $\varepsilon(h) \to 0$ as $h \to \infty$, k fixed. Since $A_0(h,k)$ converges to $\{x \in \Omega : |u_1 - u_2| < k\}$ we conclude that

$$\int_{\{|u_1-u_2|<k\}} \langle |Du_1|^{p-2} Du_1 - |Du_2|^{p-2} Du_2, Du_1 - Du_2 \rangle dx \leq 0.$$

Since this is true for all $k > 0$ we conclude that $Du_1 = Du_2$ a.e. Unless $\Omega = \mathbf{R}^N$ and $p \geq N$ we conclude that $u_1 = u_2$. This holds for all $\lambda \geq 0$. If morevoer $\lambda > 0$ then we get from (5.2) the conclusion that $u_1 = u_2$ always.

6. Continuous dependence

The third essential ingredient in any kind of generalized solution, after existence and uniqueness, is the continuous dependence of the solution obtained with respect to the data in suitable topologies. We will need this fact below when we compare our solutions with the renormalized solutions of Lions and Murat. Actually, continuous dependence comes as a consequence of the existence and uniqueness proofs. Let us consider the situation of the preceding section, but now the second member corresponding to u_i is f_i, $i = 1, 2$. Then formula (5.1) changes only in the last line which reads

(6.1) $$\int_\Omega [f_1 T_k(u_1 - T_h(u_2)) + f_2 T_k(u_2 - T_h(u_1))]\, dx.$$

Repeating the analysis we get similar conclusions for the steps (ii) and (iv), while for (iii) we get in the limit $h \to \infty$ the contribution

$$\int_\Omega (f_1 - f_2) T_k(u_1 - u_2) \, dx.$$

With this, step (v) takes the form:

(6.2)
$$\int_{\{|u_1-u_2|<k\}} \langle |Du_1|^{p-2} Du_1 - |Du_2|^{p-2} Du_2, Du_1 - Du_2 \rangle dx + \lambda \int_\Omega (u_1 - u_2) T_k(u_1 - u_2) \, dx \leq \int_\Omega (f_1 - f_2) T_k(u_1 - u_2) \, dx.$$

There are now two options, either $\lambda > 0$ or not. In the first case we get in the limit $k \to 0$

(6.3) $$\lambda \|u_1 - u_2\|_1 \leq \|f_1 - f_2\|_1$$

Theorem 6.1. *If $\lambda > 0$ the map $f \to \lambda u$ is a contraction in $L^1(\Omega)$.*

This is a well-known fact in the theory of quasilinear elliptic equations. More interesting is the case $\lambda = 0$, where to start with we have

Theorem 6.2. *The map $f \to T_k(u)$ is continuous from $L^1(\Omega)$ to $L^{p^*}(\Omega)$, $p^* = pN/(N-p)$.*

This is a consequence of the a priori bound on $DT_k(u)$ in $L^p(\Omega)$ given by Lemma 1.3, which also implies weak convergence of the gradients $\{DT_k u_n\}_n$ in $L^p(\Omega)$.

We proceed further to get a quantitative estimate of the gradient dependence. It is based on formula (6.2). For $p \geq 2$ we use the inequality

(6.4) $$(|a|^{p-1} a - |b|^{p-2} b)(a - b) \geq C(a - b)^p.$$

Recalling that $DT_k(u_1 - u_2) = D(u_1 - u_2)\chi\{|u_1 - u_2| < k\}$ we conclude from (6.2) that

(6.5) $$\|DT_k(u_1 - u_2)\|_{L^p(\Omega)}^p \leq Ck \|f_1 - f_2\|_1.$$

On the other hand, for $1 < p < 2$ the calculation is a bit more involved. For $a, b \in \mathbf{R}$ we write $A = |a|^{p-2} a$, $B = |b|^{p-2} b$, so that $a = |A|^{q-2} A$, $b = |B|^{q-2} B$, with $q = p/(p-1) > 2$. Then have

(6.6) $$(|a|^{p-1} a - |b|^{p-2} b)(a - b) \geq C|A - B|^q.$$

On the other hand,
(6.7) $$|a-b|^p \leq C(|a|+|b|)(|a|^{p-2}a - |b|^{p-2}b).$$
Combining both tricks we get on a set $K \subset \Omega$

$$\int_K |Du_1 - Du_2|^p dx \leq C \int_K (|Du_1|+|Du_2|)(|Du_1|^{p-2}Du_1 - |Du_2|^{p-2}Du_2)\, dx \leq$$

$$C \left(\int_K (|Du_1|+|Du_2|)^p dx\right)^{1/p} \cdot \left(\int_K ||Du_1|^{p-2}Du_1 - |Du_2|^{p-2}Du_2|^q\, dx\right)^{1/q} \leq$$

$$C \left(\int_K (|Du_1|+|Du_2|)^p dx\right)^{1/p} \cdot \left(\int_K \langle |Du_1|^{p-2}Du_1 - |Du_2|^{p-2}Du_2, Du_1 - Du_2\rangle dx\right).$$

It follows that
(6.8) $$\int_{\{|u_1-u_2|<k\}} |Du_1 - Du_2|^p dx \leq Ck(\|f_1\|_1 + \|f_2\|_1)^{\frac{1}{p}}(\|f_1\|_1 - \|f_2\|_1)^{\frac{1}{q}}.$$

Theorem 6.3. *The map $f \to DT_k(u)$ is continuous from $L^1(\Omega)$ to $L^p(\Omega)$.*

Let $f_n \to f$ in $L^1(\Omega)$ and let u_n, u be the corresponding solutions. On the set $A_{n,k} = \{x : |u_n| < k, |u| < k, |u_n - u| < 1\}$ the functions $DT_k(u_n) - DT_k(u)$ and $DT_1(u_n - u)$ coincide and we use estimates (6.5) and (6.8). On the remaining set the integrals we want to control are small by Lemma 1.3 and its consequences.

7. The limit case $p = N$

For this limit case much of the theory is the same. Thus, the definition of entropy solution and the uniqueness proof in the appropriate space $T_0^{1,N}(\Omega)$. This space does not imply decay at infinity. Of course for $\lambda > 0$ we have $u \in L^1(\Omega)$ which implies a certain decay in measure. Otherwise, the theory has to reflect the fact that the fundamental solution of (1.1) is a logarithm, and, consequently, solutions of (1.1) in $\mathbf{R}^N$ with, say, compactly supported $f \geq 0$ have logarithmic growth as $|x| \to \infty$. This slight difference is also reflected in the a priori estimates even for a bounded domain, the case that we will consider for simplicity, following [BPV]. We also put $\lambda = 0$. Lemma 1.1 is replaced by

Lemma 7.1. *Let be a variational solution of (1.1)-(1.2) with $p = N$ and $f \in L^1(\Omega)$. Then there exists $C = C(N,p) > 0$ such that*

(7.1) $$\frac{1}{a}\int_{\{k<|u|\leq k+a\}} |\nabla u|^N\, dx \leq M = \|f\|_1$$

for every $a > 0$. Moreover, the distribution function of u satisfies the estimate

(7.2) $$\Phi_u(k) \leq \Phi_u(k_0) e^{-\frac{k-k_0}{A_N}},$$

where $A_N = \frac{1}{N}c_N^{-\frac{1}{N-1}}\|f\|_1^{\frac{1}{N-1}}$.

This estimate will be extended to all entropy solutions by density as a consequence of the existence and uniqueness result.

We also have an a priori estimate for the gradient of the form

Lemma 7.2. *Let $p = N$ and let u be a solution as above. Then for every $h > 0$*

(7.3) $$\text{meas}\,\{|Du| > h\} \leq C(N,p)\,\|f\|^{\frac{N}{N-1}} h^{-N}(1 + \log h).$$

Using these lemmas and following the scheme of the case $p < N$ we get

Theorem 7.3. *Let Ω be bounded. For every $f \in L^1(\Omega)$ there exists a unique entropy solution $u \in T_0^{1,N}(\Omega)$ to (1.1), (1.2) with $p = N$. Morever, $u \in M^r(\Omega)$, $\forall r > 1$ and $|\nabla(u)| \in M^{N-\varepsilon}(\Omega)$, $\varepsilon > 0$.*

This theorem is proved in [BPV] where the case $p = N$ is analyzed with special attention to the fact that the space of variational solutions, $u \in W_0^{1,N}(\Omega)$ is close to the space of entropy solutions. In fact, we can prove that all entropy solutions with $f \in L^1 \log L^1(\Omega)$ are indeed variational solutions. In this case we have a more precise a priori estimate for the distribution set of u of the form

(7.4) $$\Phi_u(k) \leq c_0 \exp(-c_1 k^{\frac{N-1}{N-2}}),$$

The level-set estimate (7.4) implies an improvement the well-known *Trudinger inequality* (See [GT], page 162): If $u \in W_0^{1,N}(\Omega)$, then there exist positive constants k_1 and k_2, depending only on N, such that for any $\alpha > 0$

$$\int_\Omega e^{\alpha|u|}\,dx \leq k_1 |\Omega| \exp(k_2 \alpha^N \|\nabla u\|_N^N).$$

For solutions of (1.1)-(1.2) with $f \in L^1 \log L^1(\Omega)$ and c_1 as in (7.4) we get

(7.5) $$\int_\Omega \exp(c|u|^{\frac{N-1}{N-2}})\,dx \leq k_1(|\Omega|, N, c).$$

for all $c < c_1$. For further details we refer to [BPV].

8. Equations of general form

Following the standard usage, the analysis done in [B6] contemplates equations of the more general form

(8.1) $$-\text{div}\,(\mathbf{a}(x, Du)) = F(x, u) \quad \text{in} \quad \mathcal{D}'(\Omega).$$

The following assumptions are made on Ω, $\mathbf{a}$ and F:

(H1) Ω is an open set, not necessarily bounded, in $\mathbf{R}^N$, $N \geq 2$.

(H2) The function $\mathbf{a} : \Omega \times \mathbf{R}^N \mapsto \mathbf{R}^N$ is a Caratheodory function (continuous in ξ for a.e. x and measurable in x for every ξ) and there exist $p \in (1, N)$ and $\lambda > 0$ such that
$$\langle \mathbf{a}(x, \xi), \xi \rangle \geq \lambda |\xi|^p$$
holds for every ξ and a.e. x. There is no restriction in assuming that $\lambda = 1$.

(H3) For every ξ and $\eta \in \mathbf{R}^N$, $\xi \neq \eta$, and a.e. $x \in \Omega$ there holds
$$\langle \mathbf{a}(x, \xi) - \mathbf{a}(x, \eta), \xi - \eta \rangle > 0,$$
where $\langle , \rangle$ means scalar product in $\mathbf{R}^N$.

(H4) There exists $\Lambda \in \mathbf{R}$ such that
$$|\mathbf{a}(x, \xi)| \leq \Lambda(j(x) + |\xi|^{p-1})$$
holds for every $\xi \in \mathbf{R}^N$ with $j \in L^{p'}(\Omega)$, $p' = p/(p-1)$.

(H5) F is a Caratheodory function, continuous and nonincreasing in u for fixed x, and measurable in x for fixed u. Moreover, $F(x, 0) \in L^1(\Omega)$, and if
$$G_c(x) = \sup_{\{|u| \leq c\}} \{|F(x, u)|\},$$
then $G_c \in L^1_{loc}(\Omega)$ for every $c > 0$.

Theorems 2.5 and 2.6 are valid in this setting, as well as the a priori estimates. The approximation in the existence theorem uses the results of [Li], [LL]. Important previous work is due to [BBC], [BS], [BG1], [BG2]. Let us note that in the particular case
$$-\operatorname{div} \mathbf{a}(x, Du) + \beta(u) = f,$$
the map $f \to \beta(u)$ is a contraction in $L^1(\Omega)$, extending one of the results of Section 6. This allows to define an m-accretive operator in $L^1(\Omega)$ and generate a semigroup of contractions in $L^1(O)$ which solves a parabolic problem. Details are found in [B6].

9. Renormalized solutions

The concept of renormalized solution, introduced in [DL] in the study of the Boltzmann equations, was adapted in [B4] to equations of the form

(9.1) $$-\operatorname{div}(A\nabla u) + \operatorname{div} \Phi(u) + \lambda u = f,$$

where A is a uniformly elliptic matrix and Φ is a locally Lipschitz-continuous vector function with non-controlled growth at infinity. They take the usual $f \in H^{-1}(\Omega)$ and use the concept mainly in order to deal with the possible rapid growth of Φ. P.L. Lions and F. Murat [LM], [M], used that approach to deal with data $f \in L^1(\Omega)$, always in a bounded domain. The following definition is taken from [M].

Definition 9.1. A renormalized solution of problem (9.1) is a measurable function $u : \Omega \to \mathbf{R}$ such that:
(i) $u \in L^1(\Omega)$,
(ii) for every $k > 0$ $T_k(u) \in H_0^1(\Omega)$ and

$$\frac{1}{n^2} \int_{\{n \leq |u| \leq 2n\}} |\nabla u|^2 dx \to 0$$

as $n \to \infty$.
(iii) For every compactly supported function $\beta \in W^{2,\infty}(\mathbf{R})$ the following equation holds in the sense of distributions

$$-\mathrm{div}\,\{\beta'(u)ADu + \beta'(u)\Phi(u)\} + \beta''(u)Du \cdot (ADu + \Phi(u)) + \lambda u \beta'(u) = f\beta'(u).$$

Therefore, (formal) pre-multiplication by the 'truncature' $\beta'(u)$ is the basic tool, which can be traced back again to conservation laws, [La].

We take from [M, Th. 1.1] the following result

Theorem 9.2. *For every $f \in L^1(\Omega)$ there exists a unique renormalized solution of the homogeneous Dirichlet problem for equation (9.1).*

The concept can be naturally generalized to more general equations, as in Section 8, and the authors assert that similar results hold true.

Thus, entropy solutions and *renormalized* solutions are different approaches to the definition of a suitable generalized solution which will make the problem well-posed.

Theorem 9.3. *The concepts of entropy solution and renormailzed solution are equivalent for equation (9.1) with $\Phi = 0$.*

For smooth data there exists a unique variational solution and then the three concepts coincide. For general data we have to use continuous dependence which holds for both settings.

10. Other related works

Let us mention that the works of Rakotoson [R1], [R2] and [R3] address equations of the form $-\mathrm{div}\,a(x,u,Du) + g(x,u) = \mu$ where μ is an L^1 function or a bounded measure on Ω; he proves the existence of renormalized solutions in a certain classes $\Lambda_0^{1,p}$ with are related to the estimates (1.14); in [R3] he proves existence and uniqueness of renormalized solutions when $\mu \in L^1(\Omega)$. In all the aforementioned works the open set Ω is assumed bounded.

On the other hand, Dall'Aglio [D] studied the existence and uniqueness of solutions for equations of the form $-\Delta_p(u) + g(x,u) = f$ with $f \in L^1(\Omega)$ as limits of variational solutions, and proved uniqueness of the limit solution thus obtained.

In that sense let us say that there is a quite rich literature which sees the solution of the problem at hand in an abstract functional framework where the variational operator admits a unique well-defined closure, see Section 7 of [B6]. However, the main point of the present discussion is to characterize in clear differential terms the class where the problem is well-posed.

11. Some extensions

11.1. Maximal monotone graphs. There are a number of interesting generalizations that can be considered in the above existence and uniqueness results. One of the most common variations of equation (8.1) found in the literature concerns the possibility of including functions $F(x, u)$ which are monotone but discontinuous in u. To simplify matters, we will consider functions F of the uncoupled form

$$(11.1) \qquad F(x, u) = f(x) - \beta(u),$$

where, according to (H5) we assume that $f \in L^1(\Omega)$. We also assume that
(H6) β is maximal monotone graph in $\mathbf{R}^2$ with $0 \in \beta(0)$.

Therefore, we allow the term $\beta(u)$ to be multivalued, not necessarily defined for all values of $u \in \mathbf{R}$. The reader interested in the properties of maximal monotone graphs can consult the monograph [Br1]. This leads to the differential inclusion

$$(11.2) \qquad -\mathrm{div}\,|Du|^{p-2}Du) + \beta(u) \ni f.$$

But for the complications of taking care of the multiplicity of β, and replacing equations by inclusions, nothing essential changes in the proofs of the uniqueness result and the existence result if we assume the form (11.1) with (H1)-(H4) and the extra hypothesis (H6). Notice in particular that a complete specification of the solution involves a pair (u, w) where w is an integrable function such that $w(x) \in \beta(u(x))$ for a.e. $x \in \Omega$ and u is a solution of

$$(11.3) \qquad -\mathrm{div}\,(|Du|^{p-2}Du) = f - w$$

in the sense of Section 2. Both u and w are unique. Details can be found in [B6].

11.2. Existence for measures. Another interesting extension direction concerns the possibility of replacing the integrable functions of the second member by bounded measures. We can consider again an equation of uncoupled form, but this time we avoid the complications of dealing with graphs and take the equation

$$(11.4) \qquad -\mathrm{div}\,|Du|^{p-2}Du) + \beta(u) = f.$$

This result is proved in [B6, Th. 8.1].

Theorem 11.1. *Let $1 < p < N$, let the assumptions (H1)-(H4) hold and let $f \in \mathcal{M}_b(\Omega)$, the space of bounded measures in Ω. Assume that β be a continuous and nondecreasing real function with $\beta(0) = 0$ and assume moreover that $Domain(\beta) = \mathbf{R}$ and*

(11.5) $$\beta(\pm|x|^{-\frac{N-p}{p-1}}) \in L^1_{loc}(\mathbf{R}^N).$$

Then there exists a function $u \in T_0^{1,p}(\Omega)$ such that $w = \beta(u) \in L^1(\Omega)$ and u is a solution of $-div(|Du|^{p-2}Du)) = f - w$ in the sense of distributions in Ω. Moreover, $u \in M^{p_1}(\Omega)$ and $|Du| \in M^{p_2}(\Omega)$.

In this case we are not able to establish an entropy condition. Consequently we cannot prove uniqueness. Notice that the expressions (3.3) and (3.6) make no sense when f is just a measure, not an integrable function.

11.3. Nonlinear boundary conditions. The idea of entropy solutions has been extended to nonlinear boundary conditions equations by Mazón and collaborators, cf. [A4]. They consider the problem

$$u - \text{div}\,(\mathbf{a}(x, Du)) = f,$$

under the standard conditions (H1)-(H3) of Section 8 plus the boundary condition

$$-\frac{\partial u}{\partial \eta_a} \in \beta(u) \quad \text{on } \Omega,$$

where β is a maximal monotone graph in $\mathbf{R}^2$ with $0 \in \beta(0)$ and $\partial u/\partial \eta_a = \langle \mathbf{a}(x, Du), \eta \rangle$. A unique solution is obtained in the framework of completely accretive operators, and then the unique solution thus obtained is shown to be an entropy solution in our sense.

As in [B6]. Section 7, this leads naturally to consider the evolution equation

$$u_t = \text{div}\,(\mathbf{a}(x, Du)),$$

They add initial and mixed boundary conditions and produce results of existence and uniqueness of solutions. A similar parabolic problem for renormalized solutions has been considered by Blanchard and Murat [BM]. However, the study of parabolic problems is a different subject, worth a separate discussion.

11.4. Solutions with locally integrable data. Solutions with data $f \in L^1_{loc}(\mathbf{R}^N)$ without conditions at infinity have been considered for equations of the form

$$-\text{div}\,(|Du|^{p-2}Du) + \beta(u) = f(x)$$

under the assumption of superlinear growth on the function β, more precisely the growth of β has to be larger than the growth of the elliptic operator. Moreover, a condition like $\beta(s)s \geq 0$ is needed. Thus, the article [BGV] studies the case $\beta(u) = |u|^{s-1}u$ with

$$s > p - 1.$$

The problem is posed in $\mathbf{R}^N$. When $p_0 < p \leq N$ existence of a global solution is proved in the space $W^{1,q}_{loc}(\mathbf{R}^N)$ for every

$$1 < q < q_0 = \frac{(p-1)N}{N-1},$$

while for $p > N$ the solution is locally bounded and $|Du| \in L^p_{loc}(\mathbf{R}^N)$. Morevoer, for large s the solution has improved regularity depending on s, but always less than $W^{1,p}_{loc}(\mathbf{R}^N)$. Cf. Theorems 1-4 of [BGV]. The uniqueness of such solutions is an *open question* when $p \neq 2$. For $p = 2$ it has been solved by Brezis [Br2].

REFERENCES

[A4] F. Andreu, J.M. Mazón, S. Segura, J. Toledo, *Quasi-linear elliptic and parabolic equations in L^1 with nonlinear boundary conditions*, preprint 1994.

[Be] Ph. Bénilan, *On the p-Laplacian in L^1*, Conference at the Analysis Seminar, Dept. of Math., Univ. Wisconsin at Madison, January 1987.

[B6] Ph. Bénilan, L. Boccardo, Th. Gallouët, R. Gariepy M. Pierre, J. L. Vazquez, *An L^1 theory of existence and uniqueness of solutions of nonlinear elliptic equations*, Annali Scuola Normale Sup. Pisa, to appear.

[BBC] Ph. Bénilan, H. Brezis and M.G. Crandall, *A semilinear equation in L^1*, Ann. Scuola Norm. Sup. Pisa **2** (1975), 523-555.

[BM] D. Blanchard, F. Murat, *Renormalized solutions of non linear parabolic problems with L^1 data: existence and uniqueness*, preprint.

[BG1] L. Boccardo and Th. Gallouët, *Non linear elliptic and parabolic equations involving measure data*, Jour. Funct. Anal. **87** (1989), 149-169.

[BG2] L. Boccardo and Th. Gallouët, *Nonlinear elliptic equations with right-hand side measures*, Comm. Partial Diff. Equations **17** (1992), 641-655.

[BGV] L. Boccardo, Th. Gallouët and J.L. Vazquez, *Nonlinear elliptic equations in $\mathbf{R}^N$ without growth restrictions on the data*, Jour. Diff. Eqns. **105** (1993), 334-363.

[B4]   L. Boccardo, D. Giachetti, J.I. Diaz, F. Murat, *Existence of a solution for a weaker form of a nonlinear elliptic equation*, Recent advances in nonlinear elliptic and parabolic problems (Nancy, 1988), Pitman Res. Notes Math. Ser. 208, Longman Sc. Tech., Harlow, 1989. Pages 229-246.

[BPV] L. Boccardo, I. Peral, J.L. Vazquez, *the N-Laplacian elliptic equation. Variational versus entropy solutions*, UAM Preprint, 1995.

[Br1] H. Brezis, "Opérateurs maximaux monotones et semigroupes de contractions dans les espaces de Hilbert", North Holland, Amsterdam, 1973.

[Br2] H. Brezis, *Semilinear equations in $\mathbf{R}^N$ without conditions at infinity*, Appl. Math. Optim. **12** (1984), 271-282.

[BS]   H. Brezis and W. Strauss, *Semilinear elliptic equations in L^1*, Jour. Math. Soc. Japan **25** (1973), 565-590.

[D]   A. Dall'Aglio, *Alcuni problemi relativi alla H-convergenza di equazioni ellittiche e paraboliche quasilineari*, Doctoral Thesis, Rome, 1992.

[DL]   R.J. Di Perna, P.L. Lions, *On the Cauchy problem for Boltzman equations: Global existence and weak stability*, Ann of Math. **2, 130** (1989), 321-366.

[GT]   D. Gilbarg and N. Trudinger, "Elliptic Partial Differential Equations of Second Order", Springer, Berlin, 1983.

[Kr]   S.N. Kruzhkov, *First-order quasilinear equations in several independent variables*, Mat. USSR, Sbornik **10** (1970), 217-243.

[LU]   O.A. Ladyzhenskaya, N. Uraltseva, "Linear and quasilinear equations of elliptic type", Nauka, Moscow, 1955; English transl. Academic Press, New York, 1968.

[La]   P. Lax, *Hyperbolic systems of conservation laws, II*, Comm. Pure Applied Maths **10** (1957), 537-566.

[Li]   J.L. Lions, "Quelques méthodes de résolution des problèmes aux limites non linéaires", Dunod, Paris, 1969.

[LL]   J. Leray and J.L. Lions, *Quelques résultats de Vishik sur les problèmes elliptiques semi-linéaires par les méthodes de Minty et Browder*, Bull. Soc. Math. France **93** (1965), 97-107.

[LM]   P.L. Lions and F. Murat, *Solutions renormalisées d'équations elliptiques non linéaires*, to appear.

[M]   F. Murat, *Soluciones renormalizadas de EDP elípticas no lineales.*, Publ. Laboratoire d'Analyse Numérique, Univ. Paris 6, R 93023 (1993).

[R1]   J.M. Rakotoson, *Résolution of the critical cases for problems with L^1-data*, Asymptotic Analysis **6** (1993), 229-246.

[R2]   J.M. Rakotoson, *Generalized solutions in a new type of sets for problems with measures as data*, Diff. Int. Equations **6** (1993), 27-36.

[R3]   J.M. Rakotoson, *Uniqueness of renormalized solutions in a T-set for the L^1-data problem and the link between various formulations*, Indiana Univ. Math. J. **43** 2 (1994), 285-293.

[S1]   J. Serrin, *Local behavior of solutions of quasilinear equations*, Acta Math. **111** (1964), 247-302.

[S2]   J. Serrin, *Pathological solution of elliptic differential equations*, Ann. Scuola Norm. Sup. Pisa (1964), 385-387.

ADDRESS

Juan Luis Vázquez, Dpto. de Matemáticas,
Univ. Autónoma de Madrid, 28049 Madrid, Spain